International Perspectives on Aging

Volume 29

Series Editors

Jason L. Powell, Department of Social and Political Science,
University of Chester, Chester, UK
Sheying Chen, Department of Public Administration, Pace University,
New York, NY, USA

The study of aging is continuing to increase rapidly across multiple disciplines. This wide-ranging series on International Perspectives on Aging provides readers with much-needed comprehensive texts and critical perspectives on the latest research, policy, and practical developments. Both aging and globalization have become a reality of our times, yet a systematic effort of a global magnitude to address aging is yet to be seen. The series bridges the gaps in the literature and provides cutting-edge debate on new and traditional areas of comparative aging, all from an international perspective. More specifically, this book series on International Perspectives on Aging puts the spotlight on international and comparative studies of aging.

More information about this series at http://www.springer.com/series/8818

Matthew Lifschultz • Edward L. Schneider
Editors

Paths to Leadership in the Senior Living Industry

Springer

Editors
Matthew Lifschultz
Leonard Davis School of Gerontology
University of Southern California
Los Angeles, CA, USA

Edward L. Schneider
Leonard Davis School of Gerontology
University of Southern California
Los Angeles, CA, USA

ISSN 2197-5841 ISSN 2197-585X (electronic)
International Perspectives on Aging
ISBN 978-3-030-53968-9 ISBN 978-3-030-53966-5 (eBook)
https://doi.org/10.1007/978-3-030-53966-5

This Springer imprint is published by the registered company Springer Nature Switzerland AG
The registered company address is: Gewerbestrasse 11, 6330 Cham, Switzerland

Matthew Lifschultz • Edward L. Schneider
Editors

Paths to Leadership in the Senior Living Industry

Springer

Editors
Matthew Lifschultz
Leonard Davis School of Gerontology
University of Southern California
Los Angeles, CA, USA

Edward L. Schneider
Leonard Davis School of Gerontology
University of Southern California
Los Angeles, CA, USA

ISSN 2197-5841 ISSN 2197-585X (electronic)
International Perspectives on Aging
ISBN 978-3-030-53968-9 ISBN 978-3-030-53966-5 (eBook)
https://doi.org/10.1007/978-3-030-53966-5

This Springer imprint is published by the registered company Springer Nature Switzerland AG
The registered company address is: Gewerbestrasse 11, 6330 Cham, Switzerland

Contents

Chapter 1
Introduction

Edward L. Schneider

How to you teach leadership? That was the question I asked myself after I stepped down as Dean of the Leonard Davis School of Gerontology at USC. I had been Dean for 18 years, which is much too long. It was time to start teaching and I had been assigned to teach an existing course that introduced students to the principles of being a gerontologist. It was incredibly boring for students and for the instructor. In its place, I created a course on leadership.

Leadership has long been a passionate interest of mine. More importantly, a leadership position within the senior living industry is a logical and desirable destination for a graduate of our school. But, how do you teach leadership? I could discuss the usual texts and essays on leadership by Warren Bennis, Peter Drucker, Daniel Goleman, and the like. But I thought that our students deserved to learn leadership from real live leaders. Therefore, I decided to bring the top leaders in the field of senior living into my classroom to share their views and experiences. But how do I get these busy leaders to come to USC? There was a Belmont Village located a few blocks from my house so I made a cold call to Patricia Will, the CEO of this company. She graciously agreed to come to my class and after that it was easy to recruit others by telling them that Patricia had agreed to participate. The chapters of this book are written by many of the leaders that I found to be particularly inspiring.

Not only have my students enjoyed and benefited from the course, but in addition I have had the pleasure of getting to know these successful leaders. Many have become my friends. The course has been, and will continue to be, inspiring to my students and to me. This book collects the wisdom, knowledge, and experience of many of the outstanding leaders who have visited my class, as well as several whose careers I have admired from afar. It should be noted that the "senior living industry"

E. L. Schneider (✉)
Leonard Davis School of Gerontology, University of Southern California, Los Angeles, CA, USA
e-mail: eschneid@usc.edu

M. Lifschultz, E. L. Schneider (eds.), *Paths to Leadership in the Senior Living Industry*, International Perspectives on Aging 29,
https://doi.org/10.1007/978-3-030-53966-5_1

traditionally refers to assisted living, independent living, and memory care; here, we have also included leaders from skilled nursing and home health care, as these fields increasingly overlap with senior living.

This industry is ideal for the study of leadership for several reasons. First, the stakes are high, as the industry serves not only a vulnerable population in the twilight of their lives, but also their families, who place a sacred trust in these organizations to care for their loved ones. Second, it is exceedingly complex, combining the disciplines of real estate development and management, hospitality, and healthcare, all under the watchful eyes of regulators, the media, the public, and even trial attorneys. Third, the industry is dependent on the dedication of its frontline caregivers, who comprise 60% of the workforce and must be motivated to toil in emotionally and physically demanding conditions for low pay. To lead in this industry requires a rare combination of heart and head and the ability to achieve both margin and mission.

The senior living industry currently serves about 1.1 million U.S. seniors, a number that is expected to double by 2030. It is also a significant employer in the U.S., with approximately one million workers, a figure that is expected to grow by about 30% in the next 5–6 years. The workforce growth thereafter will accelerate, as the industry expands to meet the demand of the Baby Boomers, who in 2025 start turning eighty, a typical age for entry into a senior living community. Millions more Americans work in skilled nursing and home health care, and those numbers, too, will grow as our aging population increases. Moreover, the aging populations of China, Japan, and most Western European nations are growing rapidly. When you combine millions of workers and a growing domestic and international industry, it is obvious that there will be an increasing need for leaders. Whether you study gerontology, real estate, nursing, social work, law, finance, design, or technology, exciting opportunities in this industry lie ahead.

I believe that with this book, we have created an original approach to the study of leadership. The lessons are practical, not theoretical. Each chapter is a compressed autobiography that is personal, accessible, colorful, insightful, informative, and inspiring. These are not dry academic profiles. I promise, you will be *engaged*. You will come away from each chapter feeling like you have just met the leader who wrote it. And I hope you will want to model some or all of their practices.

Taken together, these essays present the past, present, and future of this relatively young industry. In addition, these leaders have shared their thoughts on and experiences with the COVID-19 pandemic, which, as of this writing, has been raging for over a year worldwide. The virus is especially lethal to the aged and infirm, precisely the demographic group served by the senior living industry. This tragic killer has severely strained and challenged leadership in senior living. Those who become leaders in the senior living industry may be dealing with the effects of the pandemic for years to come.

ninthbrutualmonthintheUSinplaceofsixth

Leadership is a privilege that I have been fortunate to experience. I have led as Dean of the Leonard Davis School of Gerontology at the University of Southern California; as the founder of the biomedical research programs at the National Institute on Aging; as a coach of the U.S. Naval Academy's sailing team; and as head of the medical component of the massive anti-Vietnam War protest march on Washington, DC. In addition, I have had the opportunity to learn from, work with, and befriend many extraordinary leaders, including contributors to this book.

The late Warren Bennis is widely credited with establishing the study of leadership as an academic field. Warren was a professor of business at USC for 35 years, and for much of that time I was fortunate that he was my close friend and mentor. He wrote innumerable books on leadership, including the classic *On Becoming a Leader*, which I recommend to all of you. His pithy definition of leadership is the gold standard for the discipline: "The capacity to translate vision into reality." Warren also identified the six key qualities of great leaders, which I will only mention here, as they speak for themselves: integrity; dedication; magnanimity; humility; openness; and creativity.

Leading is both an art and a science—part temperament, part character, part practice, part style. Leaders may be born, but they are also made. Leadership comes in many "flavors." That is why it is so important to observe, experience, and study many leaders. We selected 20 accomplished leaders from the senior living industry to contribute to this book. As an aspiring or developing leader yourself—whether in the senior living industry or not—you are sure to find and integrate valuable lessons from each and every one of them. I have.

Growing up, my destiny appeared secured. If not for my father dying when I was 12, I would have taken over his *schmatta*, or clothing, businesses in New York. Instead, I went to medical school, inspired by my uncle, Archie Meckler, a country doctor who made house calls, accepting whatever form of payment his patients could provide, including eggs and vegetables. Mentors and inspiring figures, like my Uncle Archie, often play prominent roles in one's development as a leader. After internship and residency, I faced—along with my fellow intern and resident, Dr. Anthony Fauci—the choice of fighting in the Vietnam War or doing research at the National Institutes of Health. I did not support the war, so I chose the NIH.

Gerontology was an emerging area of study, and after being exposed to this field, I started a laboratory within the NIH to do research on aging. At the time I did not think about it in these terms, but running a lab requires leadership. You had to set a direction for the research. You had to inspire your students and lab workers to be productive. In order to justify and continue your work, you needed results—in this case, you needed to publish quality research papers. As Bennis said, you had to translate vision into reality.

I worked under a scientific director who became increasingly jealous of the work our lab was producing. Leaders *can* have bad bosses. He spitefully limited our resources, such as microscopes, technicians, and post-doctoral fellows, making our laboratory a victim of our own success. Yet as a leader, I still had to find a way to make up for the lack of resources and get results. I was fortunate to be a member of the Johns Hopkins faculty and was able to recruit many talented Johns Hopkins undergraduates. The usual NIH laboratory had few if any undergraduate students because of their lack of experience with biomedical research. However, I had no choice as this was the only resource that was available to me. Soon, the word got out that our laboratory was a fun place to work and even better, if they worked hard, students would find their names on published research articles, which helped their applications to medical and other graduate schools. Sometimes I had a choice of writing one long paper or a few short ones. I always opted for multiple papers—not to increase my reputation, but in order to place more undergraduate names on the papers. Since we had insufficient microscopes for our work, we organized shifts of students to share our precious few microscopes. One year, we published eighteen papers in good journals—an extraordinary output, especially under suboptimal conditions.

As the Vietnam War raged, so did my opposition to it. Not only did it make no sense to me as a military history buff, but as a physician I was deeply disturbed by the sight of veterans returning without limbs and with post-traumatic stress. I joined the anti-war movement and looked for ways to advance the cause. When in 1970 the one-million-person march on Washington, DC was being organized, I volunteered to become the leader of the medical component, consisting of over 1000 doctors and nurses. I spent several days in Washington organizing food, transportation, and equipment for the volunteers and treating those who had been caught in the firing lines by tear gas. I negotiated with the mayor's staff to be able to go to the jail and free young protestors. When I arrived at the jail, a police captain threw me out. I called the mayor's office and when I returned to the jail, the captain was gone, and I was able to free over 100 very scared protestors. I was proud to do my part for a great cause. I also learned two powerful lessons: leadership can be exerted even in a non-hierarchical, if not chaotic, setting; and there are meaningful opportunities to be a leader outside of one's career.

Dr. Robert Butler, the first director of the National Institute on Aging asked me to organize the biomedical research component of the new National Institute on Aging. At this time, we needed Congress to support this fledgling institute, and we particularly wanted funding for dementia research. In the 1980s, dementia was not even considered a disease, but rather a normal part of aging. In addition, there were approximately 25 different entities that were ineffectually lobbying Congress to support research on this condition. I believed that they needed to join forces to have any effect on Congress. So, Dr. Butler, Dr. Zaven Khachaturian and I convened the leaders of these groups and persuaded them to form one large and powerful entity, the Alzheimer's Association, even if it meant their casting aside certain individual preferences—and egos—to serve the greater purpose. The whole became greater than the sum of its parts and we were able to secure funding for a network of Alzheimer's disease research centers. Today, the federal government provides $1.8 billion annually for Alzheimer's research. This was a good example of Dr. Butler's leadership in action.

As dedicated as I was to aging research, I also became obsessed with sailboat racing, a hobby that could provide aspiring leaders with an abundance of teachable moments and a way to reduce their disposable income. Just setting up for a big sailboat race requires prodigious organizational skills. Once aboard, you must captain the crew of this complicated machine, oftentimes under severe pressure, like nasty storms or tricky winds, as you make for the finish line. Sometimes, you even have to make life and death decisions. The biggest race of the year in the Chesapeake Bay, where I sailed, is the Virginia Cup. It starts in Annapolis, Maryland and finishes at Newport, Virginia. The day of the race, the wind was howling, ideal conditions for my heavy 38-foot sailboat. Five hours after the start of the race, our boat had a healthy 2-mile lead over the rest of the fleet. However, the weather forecast called for increasing gale force winds and big seas. The waves were already crashing over the bow of the boat. I love sailing in heavy weather, but there were several crew members who were not experienced in sailboat racing and had severe sea sickness. I was faced with the decision of whether to continue racing and win my first Virginia Cup and the accolades of my sailboat racing peers or drop out and be safe. Despite my visions of the glory of holding up the gold trophy, I decided to drop out of the race and not risk injury and/or death. There would be other victories to make up for this disappointment. But my first priority as a leader was the welfare of my crew. I heartily recommend this demanding and expensive but enjoyable pursuit to aspiring leaders.

My passion for this sport led me to become an offshore racing coach at the U.S. Naval Academy. It was a great honor and a privilege; plus, they took care of the upkeep and provisioning of the boats. I supervised crew safety and advised the student captains from time to time. Of course, I wanted to be behind the wheel and sometimes thought that I could have done a better job. However, my role was to bring along other leaders.

Once, while we were preparing for a race, we noticed another sailboat on a collision course with us. We were on starboard tack and therefore had the right of way. That meant that the other boat needed to turn to avoid the collision. The other boat did not turn, so our captain hailed the other boat and turned to avoid the collision. Shortly after completing his turn, the other boat turned and hit our boat. Our captain had done everything right, followed every rule of the sea, but the collision still occurred.

The Academy has no tolerance for collisions involving their boats. The Commandant summarily dismissed me as a coach and the student captain was removed from the sailing program. I was incensed by the unfairness of this decision and the consequences to the Midshipman captain. I wrote to the Commandant and pointed out that the captain had acted properly and would not ordinarily face dismissal even if the incident had occurred during wartime. I was pessimistic that this letter would have any effect. Months passed and nothing happened. Finally, out of the blue I received a concise letter directing me to report back to duty as a coach. I learned later that the Midshipman was restored to his command of a racing sailboat. As a leader, I was obligated to stick up for this student, even if it jeopardized my position with the Academy. Why? Because, simply, he had acted properly.

In 1986, I was appointed Dean of the School of Gerontology at USC, the nation's top program in this field. Though I brought to the position successful careers in bench science and directing national research programs, academia was a different world and I had the humility to know I would need help with "Deaning." Once a leader thinks they know everything, they are on the path to accomplishing nothing. I immediately set up meetings with the university's respected Deans to solicit their advice and establish ongoing relationships that I could draw on for counsel and support. Though he was not a Dean, Warren Bennis instantly became a resource for me, especially when I had to deal with serious leadership issues. I must add that I was able to pay him back by saving his life, after he had a serious heart event during one of our tennis matches.

Once I settled into being Dean, I came to realize that the job requires "herding cats." High-achieving professors all believed that they were demonstrably smarter than you and were above trifling with administrative matters. To them, my job was to obtain resources for them and then leave them alone to pursue their research. As Dean, I wanted to be sure that I was being tolerated by the faculty, since getting appreciation was unlikely. To that end, I asked my Associate Dean to take frequent anonymous surveys of the faculty for their assessments of my effectiveness. I was also fortunate that my wife, gerontologist Leah Buturain, helped me cultivate caring relationships with faculty members, especially during leadership retreats, creating a sense of shared vision. In the process, we raised substantial funds for endowed positions, eventually creating professorships and chairs for more than half of the faculty (which no other school, even Harvard, has achieved). Thus, the faculty had tangible evidence that I was working on their behalf. As the school's leader, I was a servant to my faculty, and I tended to their needs like a gardener growing exotic orchids.

In 1999, together with a very talented staff member, Maria Henke, we started the first internet-based graduate degree distance learning program in the Western U.S. I had realized that there were many potential students who worked full-time or had children or dependent elders to take care of and therefore could not come to study on our beautiful campus. At that time, most faculty and administrators were not ready to adopt this new technology, either because it was intimidating to them or because they feared it would replace them. I promised that a successful online learning program would actually make their positions more secure as it would bring additional income to USC to pay their salaries. I raised the funds necessary to start and support the program, including paying the faculty for their participation in it. We have since graduated more than 1000 gerontologists from our online programs, expanding the reach and strengthening the reputation of USC's Davis School of Gerontology.

I had the good fortune of recruiting Matthew Lifschultz to co-edit this book. Matt had a career in law, government, politics, and corporate communications before enrolling in USC's Leonard Davis School of Gerontology as a master's student. He was a student in two of my courses, and he was the best student in both. Matt exhibited a deep interest in learning and gaining knowledge, even baffling the Associate Dean for Students by requesting special permission to take more than the number of courses required to earn a degree.

When I approached potential contributors with my idea for a book on leadership in senior living, they were very supportive. However, when I asked them to write their chapters, the response was somewhat chillier. Matt proposed the problem-solving idea of interviewing the leaders, then working with them to derive their chapters from those interviews. My role was adding a few comments to each chapter. Many of our authors have commented on Matt's uncanny ability to capture and tell their stories based on their free-wheeling conversations. I believe that is a testament to Matt's exceptional communications skills, as well as his passion for the material. Matt not only substantially drafted and edited this book, but to use a Hollywood term, he also *produced* it. In summary, the book would not have been possible without the expert communications and organizational skills of Matthew Lifschultz.

I have just turned eighty and am looking forward, not backward! This book is not my last. I have just founded a company, along with two Millennial techies. While they call the company Balanced, my name for the company is "Two Geeks and a Geezer." They approached and asked me if I were interested in a company that brought fitness and mindfulness to seniors virtually. I was excited by the prospect of working with them. They offered me a consultant's position. I declined this offer and instead became a co-founder. If I am going to take part in a new company, I want to be a decision maker, not a consultant. Perhaps even a leader. Why am I starting a company when many most of my peers are in retirement communities, playing golf or bridge? Because I care about seniors who are more isolated than ever and because I have passion for my work. I believe that you need to reinvent yourself professionally every 10 or 15 years. I have consistently done this in my career, and it has kept me engaged, invigorated, and in the words of Bob Dylan, forever young.

END

Chapter 2
Thought Leadership in Senior Living

Bob Kramer

I am a social entrepreneur; senior living is the field where I have employed my talents for the last 30 years. My career in the field began when, as an elected state legislator in Maryland, I served on a taskforce to rewrite our CCRC (Continuing Care Retirement Community) regulations. I had not known what a CCRC was, but I quickly learned. Thankfully, this chance opportunity introduced me to the senior living industry.

As a Maryland state legislator, I represented the state capital, Annapolis, home of the U.S. Naval Academy. One day, a distinguished admiral walked into my office and said, "Son, my buddies and I want to build ourselves a retirement community and we need your help." I stood up and said, "Yes, sir." That meeting led to the development of a very successful CCRC named Ginger Cove, as well as my deeper involvement in the industry.

In 1991, this industry was still nascent when I co-founded and began running NIC, the National Investment Center for Seniors Housing & Care, whose mission is to advance access and choice in seniors housing by providing data, analytics, and connections that bring together investors, developers, owners, and operators. We started NIC because at that time the industry was starved for funding, and we saw that for it to grow and thrive, senior living was going to need more access to less expensive and better educated capital. Perhaps my greatest contribution to senior living as NIC's CEO for almost three decades was to attract institutional investors and other sources of capital to this emerging industry.

Attracting institutional investors to senior living was a great challenge. We had to persuade both capital investors and senior living operators that there was value in sharing their data and being transparent with one another about factors like occupancy and rental rates. By bringing these parties together, we were able to make the

B. Kramer (✉)
National Investment Center for Seniors Housing and Care, Annapolis, MD, USA
e-mail: rkramer@nic.org

M. Lifschultz, E. L. Schneider (eds.), *Paths to Leadership in the Senior Living Industry*, International Perspectives on Aging 29,
https://doi.org/10.1007/978-3-030-53966-5_2

senior living industry a recognized investment niche. However, there was considerable resistance. Many of the founders of the early successful companies, such as Holiday and Sunrise, were real pioneers—bold, risk-taking entrepreneurs—and they had found success on their own, without sharing their financial and operating secrets; so why should they now help their competitors? At the same time, investors were taking advantage of inefficiencies in the capital markets, so transparency was a threat to their competitive edge as well.

NIC quickly earned a position of leadership. My role was to motivate industry leaders to support and achieve our vision for the long-term good of the industry, even at some risk to their short-term interests. As a leader, I had to rely on *influence*, not power, because not only was NIC involvement voluntary, but these folks were actually paying my salary through their conference registrations and sponsorships. I found success by promoting the common good and by convening people so they could work together.

Persuasion and motivation were key to our success. When you don't lead hierarchically—that is, your authority is not derived from command and control—then you can exert influence only through relationships, which you must purposefully cultivate. The importance of relationships, in business and in life, cannot be overstated: when it comes to getting things done, rapport with people will always trump power and process. You simply cannot persuade or motivate people if you do not have relationships with them.

Certainly, one quality that any aspiring leader should possess—and can deliberately cultivate—is intellectual curiosity. By this, I mean curiosity about both people and ideas. Curiosity about people will help you understand, communicate with, and *connect* with them. Curiosity about ideas will help you develop, and advance, your goals. Without these capabilities, leadership and plans tend to get nowhere. Curiosity requires asking a lot of questions—of others and of yourself. And tied to asking questions is knowing what you don't know. As a legislator, for example, when lobbyists asked for my support on issues, I would ask them to make the case not only for their position, but for their opposition, too. If they could not argue both sides, I would not be interested in hearing from them again because it just is not useful to know only one side of an argument. I learned this approach from my father, who was a great influence on me. He was a law school dean and he loved a great debate. Even as a child, I had to arrive at the dinner table prepared to discuss the issues of the day. Whatever position you took, Dad would always argue the opposite side. He taught me that the more convinced you are of your position, the harder you must work to understand the other side, because you are more likely to be blind to the weaknesses of your own points and conclusions. That has always kept me humble and prevented me from making some big errors as a leader. I always ask myself, what don't I know, what could go wrong, and how will I respond?

Another great influence on me was Peter Drucker, whose work in the discipline earned him the title "the inventor of modern management." I was fortunate to have been engaged by Drucker's foundation to help popularize his classic book, *Managing a Non-Profit Organization*, and he caused me to think deeply about how one makes a social difference through management and leadership. He taught me that even, or *especially*, with a great cause, the only acceptable standard for management and leadership is excellence. Because of Drucker's teachings on mission, I insisted at NIC that we *execute* at the highest level, so that industry leaders would support us for our work and not just because it was a nice club to join. Drucker's famous dictum that "the best way to predict your future is to create it" should also be kept in mind as this industry faces significant changes and opportunities.

In addition to these two great men, I have been strongly influenced by the concept and practice of servant leadership, which is consistent not only with my religious faith but also with my belief that a real leader leads by serving and by inspiring, irrespective of his or her title. A leader must inspire others. One person with a vision is limited in what they can accomplish. Leadership means connecting people to an idea or an action that gives them meaning, purpose, and reward. The ability to coalesce people around a positive, shared vision and motivate them to achieve it—*that* is leadership.

Servant leadership *matters*, and now more than ever, as the senior living industry collectively has confronted, and will continue to confront, the unspeakable tragedy and immense challenges wrought by the coronavirus pandemic. As of this writing, it has been estimated that approximately 40% of U.S. COVID-19 fatalities have occurred in nursing homes and assisted living communities. As staff are asked to work under incredibly difficult conditions and put themselves and their families at risk, are their employers setting an example of serving them and valuing them? Servant leaders demonstrate the profound awareness that those they lead are critical to their success and they act accordingly. They understand that an essential role of management is to enable frontline workers and caregivers to do their job well in the safest possible conditions—and that means working 24/7 to ensure that staff have the personal protective equipment (PPE) and testing capabilities that they need. Such leaders think beyond the four walls of their buildings and advocate for their workers' access to food, childcare, and safe transportation to and from work. Put simply, servant leadership in a time of crisis means that staff feels their leaders "have their back." Servant leaders act with empathy and show that they both understand and *feel* their staff's concern for the residents they serve, for themselves, and their own families. Their staff therefore know that no one will fight harder to support them. When servant leaders serve their staff as family, their staff follows that example to treat the residents they care for as family as well.

Leadership in senior living is fundamentally different from most other industries because we are not merely producing widgets; we are providing a home for people at the time when they should be respected and honored both for the lives they have led and the contributions they continue to make. We make a sacred commitment to our residents, as well as to their families: *We will care for you.* As an industry, there is a societal obligation to our work. Therefore, to be a leader in senior living, one must accept that mantle of duty. Another responsibility is getting the people you lead to accept that same duty with equal reverence.

Whether it is motivating or retaining your people, culture is paramount. Practical matters like fair wages and opportunities for growth are baseline requirements, but ultimately what keeps people engaged in their work are relationships. People must feel that their company really cares for them; that they are part of a team, or even a family. Their connections with the residents must be promoted, celebrated. We know that people love relationships that give them a sense of meaning and purpose. That's our strength as an industry, and it distinguishes us from many others. To have as part of your daily work the chance to bring enormous joy and satisfaction to others is what will motivate and retain your people. A culture must be developed that is based on this.

Culture matters even more during and after a crisis, and there has been no greater crisis for the industry than the pandemic. No amount of money spent on PR campaigns or extra staff can make up for a weak organizational culture in the face of a crisis. Capital raises can come from a meeting or a call. A new strategic plan can be the result of a multi-day or multi-month effort. But strong culture takes years to instill such that staff see it, believe in it, and take confidence from it. Those organizations that emerge the strongest from the pandemic crisis will be those who had a strong culture as perceived by their staff going into the pandemic. Strong culture means that the staff, especially the frontline staff, believe that management values them and their jobs and their contributions to the organization—that the company "walks the talk" of its company values and mission statement. As noted, fair compensation is essential, but just as essential is active listening to employees and addressing their suggestions for ways to help them and the organization be better at the job they do *and* addressing the concerns they have in their daily lives, those that make coming to work ready to do their jobs well, so difficult, including day care, food security, and transportation issues. Staff do not expect management to completely solve those issues, but listening and addressing them, even partially, goes a long way to building a strong culture where staff are proud to work where they do and see their job as not just a paycheck but rather a calling and the residents as family.

Now, the particulars of culture are not prescriptive. From one organization to the next, different qualities will be more important to one than the other. However, I believe that transparency and trust are first principles for a successful culture in this industry. Without them, you cannot lead your team nor have families trust you and your organization with their loved ones. At a time of great fear and confusion, those leaders who have been transparent with families of residents, the residents themselves, staff, healthcare partners, payers, regulators, public officials, and investors will fare best. Transparency means not only sharing facts (i.e. infection rates and death rates among residents and staff) but also admitting when you don't know or when you have to admit your information is incomplete because you do not have needed access to tests and quick and reliable results. Transparency at a time of crisis is essential because a lack of it is perceived as having something to hide or even implies a presumption of guilt. Transparency means frequent, open, empathetic, and credible communication with residents, their families, staff, healthcare partners, investors, and others. There will be no going back to the relative infrequency of such communication—especially with families of residents and with staff—that we saw before the pandemic. When something bad happens in your community, you must have transparent, credible, and immediate communication. Without transparency, trust is lost, sometimes never to be regained. And trust is essential for providers of seniors housing and care, especially with families and with staff.

Another reason that culture will become even more critical as a competitive advantage in the not-so-distant future is the labor shortage, particularly for service workers, that will make even the best laid strategies impossible to execute if you cannot attract and keep great people. The dramatic decline in the availability of informal caregivers cannot be exaggerated. Currently, more than two-thirds of all long-term care in this country is provided informally—meaning unpaid—by spouses, significant others, children, and grandchildren. According to AARP, the current ratio of caregivers (between the ages of 45–64) to cared-for (those over 80) is 7–1, but by 2030, it will fall to 4–1, and by 2050 it will be 3–1. That, exacerbated by the paid-labor shortage, means there will be huge demand for the services provided by the senior living industry, but it also means entirely new solutions will need to be imagined and executed, because organizations won't be able to simply throw people at the problem. Many of the industry's top leaders already recognize that corporate culture in senior living is more than a set of beliefs; it's an investment that an organization makes in itself.

Along with a pronounced labor shortage will come those changes brought by the Baby Boomers, who will do nothing less than redefine aging and retirement. They will exhibit new behaviors around work and consumer consumption, by retiring later and later, or not at all, and by demanding a lifestyle that reflects *their* needs and desires. Just as they collectively said in their youth, "Hell no, we won't go" (to the Vietnam War), aging Boomers will largely reject being told how they must live, whether by their families or the culture at large.

The old view of retirement and of aging was about the "D-word." It started with "**d**eclinist," which described the general view of aging. Until now, retirement meant you **d**isengaged, you **d**isconnected, and you became **d**ependent until your **d**isappearance, or **d**eath.

The increasingly dated promises of our industry were safety, security, and comfort. These defined the senior communities that served the Greatest Generation, which went through World War II. They could be called "age ghettos" because they were typically filled only with older people and were often segregated, physically and socially, from the rest of society. Senior living communities in the future will be integrated with their surrounding neighbors, neighborhoods, communities, or else Boomers will not want anything to do with them. Instead, the old model inevitably is being replaced by an E-word model: it will be based on **e**mpowerment, and with it will come **e**ngagement, **e**nrichment, **e**xperience, and **e**njoyment.

To borrow a phrase from Marc Freedman, the founder of Encore.org, which helps people 50+ find meaning in new careers, retirement for Boomers will be one-part purpose, one-part passion, and one-part paycheck, in varying combinations, depending upon their individual position in life. Some will have to work, and the paycheck will be a necessity. Others will choose to work, out of a sense of purpose, perhaps to make a difference in people's lives, such as by teaching, mentoring, caregiving. For others, work will be driven by passion; "I've always wanted to do X, Y, or Z." That means 10 years from now, one-third of our senior living residents will still be working; in 20 years, more than half will.

Economic and demographic factors will lead us into at least a decade of what the late Harvard Business School professor Clay Christenson famously called disruptive innovation. This describes a process whereby newer, smaller companies with fewer resources will be able to successfully challenge established incumbent businesses that are, one might say, too busy focusing on delivering their products and services to those that appear to be their most reliable customers, thereby ignoring the needs of other prospects. Disruptors will find ways and reasons to deliver what Christenson called "more-suitable functionality" and frequently, at lower prices. Incumbents, having lulled themselves into believing in their inherent and permanent market superiority, will tend not to respond vigorously. But new entrants will deliver the product and performance that the incumbents' mainstream customers require, while preserving the advantages that drove their early success, like originality, creativity, and scrappiness. This process will come to define the senior living industry, and wherever you are in your career arc, you should take note of this.

Moreover, the process of change in senior living will involve *continuous innovation*, meaning doing what you do better. Success will not come to those who are stagnant as the demographic wave washes over the industry. Improving your products and services will have to become routine. The good news is, doing continuous innovation is great training for doing disruptive innovation, because it teaches you which systems need disrupting. You will need to learn how to think about the customer in new ways, like Netflix did in destroying Blockbuster. A surfing analogy is apt, since I grew up loving to surf. If you paddle into the wave too soon, you wipe out spectacularly; if you paddle in too late, you miss the wave and get marooned. Which, in the senior living industry means, you go out of business. Bill Gates says we tend to overestimate the change that will happen in the next 2 years and underestimate the change that will happen in the next ten. That applies to our industry. Enormous change is coming, but it will not all happen overnight.

Part of the coming change will involve access and affordability, and this is an issue that our industry, and indeed our nation, cannot ignore. Private-pay senior living is primarily focused on the top third of our society in terms of financial resources. The economically disadvantaged are served, at least in part, by government-funded safety net programs. Access for middle-class Americans has traditionally come from combining income, a calculated, slow spend-down of assets, and additional financial help from adult children. This existing "system" cannot be relied on to provide options for all. Indeed, according to a landmark analysis recently conducted by NIC, along with researchers at NORC at the University of Chicago, Harvard Medical School, and the University of Maryland School of Medicine, slightly less than half of all middle-income older adults will be able to afford to move into a senior living community by the year 2029. That is the case for several reasons: 1) demand could be so great that many more units will be needed for this middle market; 2) the Boomers have saved less than they needed to; and 3) we're living much longer. These factors will conspire to create massive need and massive demand in a population that will double in size in the next 10 years, to over 14 million. Of course, this means there will be enormous, ongoing opportunities for the industry, though we also must recognize that even if traditional investors and operators get focused on developing such a product for middle-income older Americans, the needs of this cohort will not be adequately met. To do so will require meaningful public sector involvement, such as tax breaks and tax incentives, as well as increased participation from investors, ranging from pension funds and family offices to health insurance and life insurance companies.

It should be noted that the real estate piece is the easier part of this problem. The far harder challenge is the care piece. People may be satisfied with modest accommodations, but not with poor care. Technology and telemedicine must play a role in delivering high quality care in modest settings, because care itself is quite expensive. It would be a national disgrace if we fail to confront and solve this issue. Today many seniors spend down their assets to qualify for Medicaid in order to utilize subsidized facilities. This overstress to the Medicaid system affects the truly poor for whom Medicaid was designed. It also causes widespread shame and anxiety for many middle-income Boomers by forcing them to accept a condition that they will regard as a major failure, being forced to go on the government dole after a lifetime of work. As a new generation of leaders enters senior living, I would admonish them, with equal parts of hope and urgency: Big problems necessitate bold solutions.

As for bringing in ideas, we should look to Japan, which also has a large aging population and a dearth of care workers. They offer two policies that would be prescient for us. First, the government has made it a social norm that when you retire from your first career, your second career will be as a caregiver, so they are training and paying caregivers in the 60- to 80-year-old cohort to care for people from 80 to 120; yes, 120! Second, they are channeling funds and efforts toward developing technologies to support these caregivers, notably in exoskeletons or body suits and in robotics, including social robotics for companionship.

Speaking about 100-year-olds and robots may sound futuristic, but that future is just about now, and tomorrow's leaders will have to lead in transformative cultures. For operators, the workforce challenge will be, how can you be employer of choice in your *market*, not just in your field. You will have to create a culture that promotes great relationships as well as a sense of meaning and purpose. This does not happen overnight, so operators ought to be starting on this *now*. The workforce challenges are tough now, but they are going to get even tougher. Culture will be the defining issue of the industry in the not-too-distant future. One major obstacle is that in our industry you are preoccupied with day-to-day operations—care, cost of capital, occupancy—which are all-consuming. But those who do not create the culture that attracts and retains people will fail regardless of how brilliant or original their strategic plans.

In a people-centered, people-driven industry, it simply cannot be overstated that *the* most important decisions leaders make involve people, people, and people. As the famous line from Peter Drucker goes, culture eats strategy for breakfast. If your people aren't passionate about your mission, failure is all but guaranteed. The profound truth of this observation will be even more resonant over the next 10 years.

Strategy, of course, is also important, and as a leader it is your responsibility to always be learning, reading, listening, and anticipating the future. Our resistance to change can be insidious, and many get so caught up in daily or quarterly performance numbers that they lose sight of the big picture and fail to prepare for the future. A leader in senior living must always be probing, asking questions such as: What are the lifestyle trends? Where is the overall sector headed? How is the customer changing? How are financing models developing? How will disruption of the healthcare payment and delivery system affect our residents? How do we leverage technology? What do the data tell us? Sound strategy emerges only from deep questioning.

Communicating your culture actually starts with feedback coming up to you as a leader, not by sending it down from above. It is imperative to seek feedback constantly, especially in anonymous and candid form. This is a *sine qua non* when it comes to whether or not the organization's activities are aligned with its core values. As a leader, your own perspective is bound to be flawed and prejudged. One must break out of that self-contained feedback loop for objectivity, which can and ought to come from your own employees at all levels, as well as board members and key customers. And ideally, the feedback is continuous.

Planning for the future and casting a vision also matters greatly—especially when dealing with the effects of an overwhelming crisis. In a recent essay on the COVID-19 crisis, published in the *Wall Street* Journal, Henry Kissinger wrote that "the historic challenge for leaders is to manage the crisis while building the future." He went on to observe that "the crisis effort, however vast and necessary, must not crowd out the management task of leading a parallel enterprise for the transition to the post-COVID order." For leaders in seniors housing and care, this means that, even while responding to an unprecedented crisis in which their residents are at highest risk, they must also anticipate and plan for what the new reality will look like in senior living as a result of the pandemic. What will change in the expectations of residents and their families and of staff? How will that impact existing business models? How will the huge shift to embracing the delivery of healthcare and management of health conditions at home impact the routine practice of sending residents to the hospital or doctor's office for their healthcare? In addition to planning for the future, leaders in seniors housing and care must also cast an attractive, aspirational vision for their congregate setting and its benefits in the light of public concern and negative publicity about the health risks of such a setting. In any crisis, including the COVID-19, those leaders who anticipate and plan for the future, and are prepared to inspire and rally others around a positive vision for that future, will stand out. The inverse is equally true: those leaders who fail to plan beyond the immediate moment will find their organizations left behind as the rest of the industry inevitably responds to it.

Pride, overconfidence, and a lack of humility are fatal flaws. If these flaws exist, organizations fail, whether gradually or suddenly. Many ethnic cultures have a version of the expression, "a rotting fish stinks from the head down" because hubris has no borders. When leaders believe their model is perfect, or even better than the rest, beware—as an employee, an investor, or even as a resident. Because that is when leaders stop listening and learning. Quality cannot improve under such a regime.

An example of this for me personally is, in my career I have done *lots* of public speaking, and I think that I am pretty good at it. But I do sometimes come up short of my own expectations, to say nothing of the audience's. And, not surprisingly, the only times I have really fallen flat on my face is when I did *not* have the pre-speech butterflies in my stomach. The cause of my failure? Overconfidence! When I'm nervous before I speak, I know it will go well, because I'm humble and focused on my audience. Intel's visionary co-founder Andy Grove said, "Business success contains the seeds of its own destruction. Success breeds complacency. Complacency breeds failure. Only the paranoid survive." He meant, do not rest on your laurels. I would say that only the humble continue to thrive.

If you desire to become a better leader, read. Read widely and curiously. Among my favorite suggestions are Max Dupree's *The Art of Leadership*, almost every book Peter Drucker wrote, and the many fine books on servant leadership, including those by Robert Greenleaf. In addition, an aspiring leader ought to *engage* widely, especially on matters of which you feel certain. For example, find people with whom you disagree and invite them to challenge your assumptions and your conclusions. As an elected official, I used to gather groups of folks who did not support me on key issues, and hold sessions to exchange ideas, points of view. I always learned a lot, and not only about the subject itself, but also about those people, and even about entirely separate issues. Leaders of organizations can—must—do the same.

Early in my career, I received the gift of a profound insight from a colleague/mentor. He explained that lasting success ultimately would come to me not due to the things I was good at, but from a self-awareness of my greatest flaws as a person and as a leader. Because those flaws, which every one of us have, are, if uncorrected, what will eventually take you down. The better you are at understanding and addressing your own weaknesses, the better and long-lasting a leader you can become.

And, by the way, that is pretty good advice for relationships in your personal life, too.

END

Bob Kramer is Founder and Strategic Advisor at the National Investment Center for Seniors Housing & Care (NIC). He is also the President of Nexus Insights a thought leadership platform focused on innovative ideas and actionable models in housing and aging services. He is broadly recognized as one of senior living's most influential and high profile thought leaders and connectors. With over 35 years of industry leadership, he has earned the reputation of "agent provocateur" in the seniors housing and care industry. Mr. Kramer has been described as an "ice-cutter" and scout in identifying industries and trends that will disrupt the future of seniors housing, aging services and aging more broadly.

In 1991, Mr. Kramer founded NIC to advance access and choice in seniors housing and care by attracting capital to grow the sector. Under his leadership, NIC became the go-to resource for data and analytics for the rapidly expanding seniors housing and care industry and an acknowledged thought leader for new ideas and trends impacting the provision of housing, services and care for older adults. He stepped out of the CEO role in 2017 and continues to serve NIC in the role of Strategic Advisor.

In 2019, Mr. Kramer conceived of and championed "The Forgotten Middle," a landmark study of the health and socioeconomic status of middle-income adults who will be 75+ in 2029 and their ability to afford private pay senior housing and care. The report on the study, published in Health Affairs, was the 5th most read article appearing in this leading health policy monthly journal in 2019.

Mr. Kramer was educated at Harvard and Oxford Universities and holds a Master of Divinity degree from Westminster Theological Seminary.

Chapter 3
Senior Living Visionary

Keren Brown Wilson

I was born in West Virginia and raised in eastern Tennessee, where my parents' families had lived for 200 years. Isolated and poor, family and church were central to most peoples' lives. My mother was widowed when she was young and had four children still at home to feed, clothe, and send to school by herself. Life there was hard. When it came time for college, I chose to go to the furthest and most urban place I could find. I graduated from the University of Washington, in Seattle, writing an undergraduate thesis on aging.

As a teenager seeking independence, I was not happy my mother and my younger sister decided to follow me to Washington. However, it turned out to be a blessing in disguise, because Mom soon had a stroke that caused significant impairment, and she lived for the next 10 years in a nursing home. Being nearby for much of the time, I was able to assist her.

I was always interested in how one can help others live their lives to the fullest. While attending UW I studied psychology and worked in aging services. I also developed an interest in program development and public policy. Then I continued to work with older adults in community-based care while studying for a Master's in Public Administration, with a focus on aging policy. While I was studying for my Master's degree at Seattle University, I dreamed that someday I would join a presidential cabinet as the Secretary of the U.S. Department of Health and Human Services so I could make policy at the highest level.

When I finished my Master's, I took a job with Washington State's Division of Research, evaluating aging service outcomes and conducting one of the early studies on case management. I also met my husband and he encouraged me to return to school to get a PhD. That was one way to examine both theory and practice, and I was keen on combining the two disciplines. With a fellowship from the federal

K. B. Wilson (✉)
JFR Foundation, Clackamas, OR, USA
e-mail: kwilson@jfrfoundation.org

M. Lifschultz, E. L. Schneider (eds.), *Paths to Leadership in the Senior Living Industry*, International Perspectives on Aging 29,
https://doi.org/10.1007/978-3-030-53966-5_3

Administration on Aging, I pursued my PhD at Portland State University and ultimately wrote my disseration on divorce in late life. At the outset of the program, I told my mother that I was going to get a doctorate so I could conduct research and teach in the field of aging. Fatefully, she responded, "Why don't you do something to help people like me?"

It was an arresting comment, and it changed my life. It set me on a career path of real world problem-solving instead of academic study. Having witnessed how she lived in the confines of her nursing home I knew what she meant when she elaborated, "I want to go to sleep when I want. I want to eat what I want. I want to have my cat and my own furniture. I want to be able to lock my door. I want to *live*, and I need help so I can have a *life*." She was describing the classic failings of nursing home living as a long-term care option, which were well known at that time: grim, institutional settings providing little or no autonomy for the residents. In that moment, my mother vividly framed the great need for assisted living, which at that time did not exist. For many infirm elderly, nursing homes were the only option, especially for those dependent upon Medicaid to pay for their care. These fundamental human desires informed my initial view of what living arrangements for older adults in need of care should be. And my career arose from the simple act of carefully listening to her. That itself provides one of the key lessons for leaders and aspiring leaders alike: *listen carefully*.

Her comments focused my interest in policy, systems, and programs in the area of long-term care and, specifically, how such care needed to be changed. Her death during the first year of my doctoral studies only strengthened my resolve to help bring about such change.

Gerontologist M. Powell Lawton had an enormous influence on me. His theory of "Environmental Press" propounded that vulnerable older adults would either thrive or fail depending upon whether or not they lived in an environment that was optimally suited for their *independence*. It was an enlightening idea: how could we create an environment that made it possible and worthwhile for people like my mother to be more independent, while also recognizing the need to be assisted by caregivers? I believed that getting these elements properly balanced would help achieve the elusive goal of healthy independence. It was Lawton's theoretical framework meeting the real-life context of my mother's plea for a life of autonomy and control that was the genesis for assisted living as I conceived it.

While pursuing my doctorate, I visited a number of different venues of care, including many independent housing facilities. From my observations, I came to believe that it would be good public policy for the State to add services to independent housing, enabling residents to stay there longer and live better. But when I introduced this concept, it was met with stiff resistance from providers of both senior housing and senior services. At this time, I was working for a non-profit trade association as the Director of Education and pitched the idea to everyone I knew. There were no takers.

Apparently, it is not my in my DNA to back down from opposition. An acquaintance from PSU came to me wanting to build a nursing home. I told him he could never get a certificate of need but that there was a real market for something else.

Along with my PhD husband, who also spent his professional life working with older adults, we set out to find the money. A private investor with a construction company provided bridge financing for a majority ownership. Thus, we had a developer, a market feasibility study, a builder, and me. My job was to figure out how to make ideas into reality operationally. Mind you, I was still working full time at this point. This was an early lesson. Somebody always has to be all in. If that person happens to be you then *you* become consumed. And it is always a good idea to have some sort of plan to survive financially. My plan was having a husband who worked, plus a lifetime of thrift and delayed gratification.

The prospects looked good until we applied for State licensure and were denied. We were told, "This does not conform to 'residential care' and we do not have a category called 'assisted living.'" When I demanded that State officials cite any rules stating that our building could not be licensed, they could not do so. Because there were no such rules.

I challenged the State, telling the powers that be, you must give me a license or else we will operate without one. We negotiated and compromised, but I never abandoned my determination to provide this much-needed form of housing for older adults. Indeed, a year later we received a waiver to operate with the designation "special residential care," and we were in business.

That first assisted living center consisted of private apartments with locking doors, separate heating and cooling, and kitchenettes. It had common spaces for dining and laundry, and a dedicated space for personal care and nursing services. This may sound run-of-the-mill today, but this was in 1983, and our offerings were new, if not *radical*. Called Park Place, it was unlike any conventional nursing home. The market was very receptive, and occupancy quickly reached 100%, despite being all private pay, with no government subsidies. To my surprise, the "people problem" we faced was not in attracting residents; it was in hiring qualified staff. Before the building opened, I could not find anyone to run it. All the licensed administrators we attempted to recruit thought we would fail, and they did not want to be associated with this risky experiment. Therefore, I had to run it myself, despite having no management experience.

Not long after we opened, we began planning our second facility that poorer folks like my mom could afford. Her words were, "Why don't you do something to help people *like me*?" and my mother could not have afforded to live in our first building. I wanted the next building to demonstrate we could secure subsidies and provide housing for people like her.

We asked the State to approve a Medicaid demonstration, or pilot, project. I invited officials to tour our facilities and proposed a contract to house Medicaid clients in 20 of the 127 units. To get the State's buy-in, I suggested that they choose our clients and conduct their own outcome evaluations. I was confident in our approach, so I was perfectly willing to allow State officials to have say-so over selecting the residents, whose housing and services they were effectively funding. After building, opening and filling two large buildings in 5 years, I was ready to strike out on my own. I cashed in my pension which had about $5000 in it and opened a one room office with a phone and a computer. My husband and I formed a new partnership and it took two more years to open our first small building. Sweat equity is how our new partnership came to be. But I didn't realize that in replication, the "sweat" contribution would become less for other partners while mine always remained high. It is a failure that often plagues "true believers" who really want something to happen. It is a decision I have made more than once—being willing to give more, or to agree to conditions that I wouldn't normally—because I wanted something to happen so badly. We opened our first small building in order to achieve my ultimate goal, which was to provide our type of housing in smaller, more rural communities. I believed that smaller buildings made it easier for the residents and to be in places others would not build. That approach became our national model, and that first 25-unit building is still operating today and continues to be revenue positive.

Having established myself as a pioneer in housing for seniors, I soon enjoyed many opportunities to evangelize this new concept of assisted living. I wrote articles, gave speeches, and participated in conferences all over the U.S., talking to governments, trade associations, and universities; I would talk to anybody who would listen. We promoted and hosted tours, determined to show people that our new approach worked. At around the same time, Paul Klaassen, the founder of Sunrise Senior Living, was developing his new model of assisted living on the East Coast, so momentum for this nascent industry was growing.

But, being a new industry, there was a leadership vacuum. Though I did not set out to become a leader of a movement or even of a company, I found myself in both of these roles. I recognized the enormous responsibility and opportunity that I had taken on, of helping to bring a better life to countless older adults in America, and I didn't stop to take a breath for about 20 years. In under a decade, I went from running 6 privately owned facilities in Oregon to heading a publicly traded company with 183 facilities and 3000 employees across 18 states. I wanted assisted living to be widely available in all 50 states, including to Medicaid clients. Indeed, one of my greatest failures and regrets is that I was not able to achieve that goal. I also wanted it to be "perfect," but I learned the hard way that is nigh impossible when an organization grows beyond the span of control of an individual. People are not perfect, and their needs and goals are not always aligned with yours. Fortunately, though, interest in assisted living boomed, investment capital followed, and so did tremendous growth. I became one of the public figures of our industry, and my expertise and experience were frequently called on, whether by fellow operators, policymakers and regulators, or the media.

My role as a leader in housing for seniors did not end there. In 2001, I created the Jessie F. Richardson Foundation—named for my mother—to develop sustainable, replicable models of elder care in regions with limited resources. I was influenced to focus that work in Central America by Dr. Martha Pelaez, who was an advisor on aging to the Pan American Health Organization, part of the WHO. She had requested my help on housing and long-term care issues and when we met in her office, I glimpsed a photograph of an older woman sleeping in a wheelbarrow. I asked for an explanation, and Dr. Pelaez replied, "That's the face of aging in most of the developing world." I was dismayed but motivated and immediately thought, "I must *do something* about this."

When I went to the big players in international development to request funding for a project to help the aging poor in Latin America, I was told repeatedly some version of either, "That won't work" or "That's not our issue." However, I was not going to take "no" for an answer, especially after visiting Nicaragua and seeing that many residents of its *hogares de añcianos*, or nursing homes, slept on bare wooden pallets and had little food or medicine. I was determined to build a collaboration to bring help to Nicaragua's poor elderly.

To attract support from potential partners and funders, I needed a compelling proposal for how we could help and found it in the approach called Asset-based Community Development (ABCD). ABCD drives change and development through citizen involvement, drawing upon the skills of local people to address community issues. It uses outside assistance but takes its agenda from those affected. While it had been used for economic development, I wanted to apply it also to social development. I spent several years explaining and pitching—and listening—in order to convince folks that an outside organization could export its expertise to help build community to support older people. Ultimately, I found backers.

As a result, the Foundation has been able to help change living conditions for thousands of older adults by providing health care, building care capacity, and fostering economic opportunity. We helped establish Nicaragua's first National Council on Aging and trained 1000 care workers, students, and volunteers in 20 cities. In addition, we developed and hired the country's first geriatric physician, who treated thousands of elders and created a system of community-based navigators who develop local community support for the *hogares*. We created 35 Spanish language videos to train caregivers and an administrator training program—all of which is delivered remotely.

With about 40 years of leadership in the field of senior living behind me, when perhaps some would have said, "my work is done," I recently chaired the launch of a non-profit called Age +, which is a public charity dedicated to developing non-subsidized affordable housing and services for low-income older adults in the rural communities of Oregon, my adopted home. This new effort was sparked by the physician and community health expert Dr. Tina Castañares, who advocated for the unmet needs of the poor older population of rural Oregon. They often face conditions of inadequate access to care and personal resources and when I examined their plight, I knew that I was called, yet again, to *do something.*

With government subsidies for senior housing diminishing just as the need for affordable housing and services is rapidly increasing, Age + is producing a modular housing design that will rent for between $500 and $600 a month, providing poor, rural older adults access to relatively inexpensive housing. Few if any developers in the U.S. are attempting to provide non-subsidized affordable housing and services, but if it does not start happening soon on a large scale, there will be a huge increase in the number of homeless older adults—and not just in rural Oregon, but nationwide. The sense that, if we don't do something, no one will, can be a powerful motivator. So, in addition, Age + has launched "Circles of Care", which is a community-based volunteer support system for older adults who lack the money to pay for care or support. Given what we know is coming in the way of an increased older adult population with decreasing personal assets and less capacity of government to meet increasing needs, it seems prudent to think outside the box once again to do something.

In each of the three phases of my career, I have forged a similar path to a leadership role. What has led me on that path is the interplay of three factors: first, my core belief that all older individuals deserve a quality of life with access to support, services, and adequate housing; second, my personal philosophy, borne of observation and experience, that leading a change effort requires collaboration; and, third, my natural temperament to identify and surmount seemingly intractable problems—in short, to *do something*. And I believe that anyone who aspires to leadership in the field of senior living can attain such a role by adopting a similar mindset.

3.1 Core Belief

From the moment I began to build my career, I had the good fortune to feel passion for it. While studying aging in college and graduate school, I also had jobs in the field and was highly gratified by my work. Helping to care for my mother exposed me to the dismal state of the existing nursing home industry and stimulated my desire to improve if not eliminate it. As a result of my study, work, and personal experience, I developed the core belief that all older adults deserve a decent quality of life as well as a persistent belief that I can be a part of changing how things are done in long-term care. Housing and services for seniors became my *passion.* Such passion, of course, will cause you to work with determination, open your mind to new ideas, and bring you joy when you accomplish your goals. My passion has been the driver of every endeavor in my career, and it has enabled me to innovate, overcome obstacles, achieve objectives, and love my work. Without such passion, one will tend to respond mechanically to her work's challenges and responsibilities.

In our industry, the stakes are people's lives and the domain is exceedingly complex and demanding, so passion for the work ought to be regarded as a *prerequisite* to leadership. Possessing the core belief that older people must have access and choice in housing is a good foundation for all those aspiring to leadership in this industry.

3.2 Collaboration

Had I stayed in academia after earning my PhD—teaching, researching, writing—much of my work would have been conducted solo, with the occasional co-authored research paper. Once I began the real-world push for the assisted living model, though, I needed partners and allies, from the individual to the institutional. However, this was not readily apparent to me.

For a time, I enjoyed singlehandedly applying my clear vision and competitive will to developing assisted living. Indeed, when others began to execute similar concepts, I had some apprehension; would they be faithful to my concept? Wondering if I was perhaps being overprotective, I shared my concern with one of my mentors, Dick Ladd, who directed Oregon's aging services. He was blunt: "Would you rather it be perfect for a few people or be better for a lot of people?"

It was a revelatory moment, exactly what I needed to hear. It got me to recognize what I call "gatherer leadership," and I henceforth saw myself above all as a *recruiter*, seeking to personify that age-old expression "there's strength in numbers." The circumstances of trying to change an industry, or perhaps even create a new one, necessitated that others be a part of it, and bringing those people together would become my priority. I recognized that, while our culture often mythologizes singular leaders, a movement cannot occur with just one person. Collective force is far more likely to create real change. To build a movement, or even a project, it is mission critical to have a cadre of people sharing the vision, even if they might implement that vision somewhat differently.

I set out to learn all I could by listening to and watching others, while also sharing everything I knew with anyone who asked. In my own organization, for example, I worked shifts on the floor, and asked my staff, "Why do we do that and how could we make it easier and better?" Every conversation, meeting, or conference was a collaboration on assisted living. And this illuminated the point that, to attract the people the mission would require, I would have to be in the field *with* them, not out in front of them; I thought of it as leading from behind.

Additionally, I was fortunate to view competition not as a process through which I must win, and you must lose, but instead as potentially a perpetual cycle in which participants win merely by doing their work better and better. So, if I compete against myself, you do the same, *and* we collaborate, everyone ends up better off. Competition conducted this way is a form of collaboration, as all the participants seek to do better and, in the process, improve the outcome for all. That is why I have always been willing to help even those who were ostensibly my competitors, despite the fact that I am very competitive—though principally with myself.

Another opportunity to embrace collaboration comes from recognizing that not only is an issue *important to* you, and is thus worth your greatest efforts to effect change, but also that it is *bigger than* you, and will thus benefit from the contributions of others. My path to leadership has always been characterized by interactions with other people who were trying to do similar things, and the sharing of ideas, support, and resources.

Ideally, your strategy is reached collaboratively. You have an idea, and you talk about it, from the bottom up and from the top down, with your board, employees, and partners. Though they don't dictate your moves, their perspectives do inform the final product. That product is always going to be better when you engage in a collaborative process.

3.3 Do Something

Indeed, the most important role of a leader is to set strategic direction, then identify opportunities that are aligned with that direction and build a framework for how to move in that direction, including finding the people who can help. Leadership is action-oriented, with actions being designed to effect change.

That is why throughout my career there has been such resonance to my mother's plea to "do something to help people like me." Whether she meant to or not, she was imploring me to be a leader, to help lead change to a system that was not working for many of the people it served. Now, had I been of a different temperament, her words may have fallen flat with me, and I might have proceeded to pursue purely academic endeavors. But I have always been drawn to try to fix stubborn, intractable problems, and her words ended up launching me on a very different career trajectory, on a path toward leadership. Thus, the words *do something* were very much on my mind when I confronted the plight of Nicaragua's and rural Oregon's under-resourced older adults.

Studying and observing other leaders can yield valuable lessons, particularly how to avoid making mistakes. I am always learning from my own mistakes, as well as those of others.

Perhaps the mistake most commonly made by a leader is believing that you know more than you actually do and that you can do most things as well as or better than anybody else in your organization. Sometimes success makes us feel overly important and infallible—we begin to believe our own press. That lack of humility can be a really big problem. It undermines the trust that leaders should be cultivating for their people. This is often aggravated by the presence of employees who believe it is in their best interest to say what they think the boss wants to hear. That only reinforces the ignorance of the boss. As a leader you must have the confidence in yourself to have people around you who possess knowledge and skills that you don't and be willing to let them do their jobs. Leaders must be able to get out of the way! Top-down structures are especially inimical to the work of senior living, where most staff is on the front lines, not headquarters, and need to be empowered to make decisions and adjust policies without a lot of red tape.

Leaders also too often commit what I call "the assumption error," which is assuming that everybody is on board with their policies, decisions, and directions. We assume that everyone—our staff, the community, the authorities—understands, cares and agrees. Such assumptions increase the risk both of individual and organization failures. How can you build a coalition, how can you collaborate, when the parties involved do not all agree, or at least find those areas where they agree to disagree? Fortunately, there are correctives for the assumption error. First, you need to understand what is driving you and the people that you work with. We all have different motivations, and you cannot ascribe your motivations to others. When you know others' motivations, then it's easier to find ways to work together. In addition, rigorous attention must be paid to your internal communications. You must overcommunicate, and be simple, direct and transparent. And then be sure your communications are observable in action.

Finally, I would caution aspiring leaders to look out for what I call the "A & F Syndrome," which means we tend to pay the most attention to our grade "A" and "F" employees, promoting the former and marginalizing the latter. But most employees are neither, so one must work hard to find ways to help *them*—likely 80% of your staff—to be successful. I also call this "How to Raise Your Performance Mean," and especially in an industry where the work of virtually every employee can impact peoples' lives, it is a problem, and an opportunity, that leaders can't ignore.

We did not need a global pandemic that preys on the older population to demonstrate that our work impacts peoples' lives, but COVID-19 certainly accomplished this. The pandemic presented a great challenge to leaders of all kinds, especially, as it turns out, to those in the field of senior care. When in February 2020 a reporter asked me about the risk to older adults in care settings, it seemed to have come out of nowhere given that homes generally had pretty good policies and procedures around infection control. But the coronavirus pandemic revealed how unprepared the United States actually was at every level—national, state, local and, yes, even at senior care settings. Some things were startling, such as the disparate death rates among communities of color or the shortages of PPE. In the midst of the pandemic, the public broadened its view of who were the real heroes in communities, at least temporarily, to include caregivers of all kinds, environmental services providers, the men and women who stock shelves in grocery stores.

While leadership, especially at the federal level, fell short in many ways, strong leadership was shown by many in senior care field. By definition, leadership is about recognizing risks, whether the risks are financial or from rampant infection, and envisioning a response. Communication skills that are always important with residents and their families, staff and the public became especially critical during the pandemic. Lack of adequate and transparent communication frightened many, leading to heightened anxiety and rampant speculation about what was happening. Without excellent communication in situations like the coronavirus, a leader's and an organization's credibility is at risk, and when lost, that is very difficult to regain. Paradoxically, even if news is uncertain or bad, facts can be comforting if presented with empathy.

On a practical level, during the coronavirus outbreak, the most effective leadership traits are, be early, be fast, be all in, and finally, be transparent. Senior care settings that didn't wait for the state to issue directives about closing their buildings to visitors or to offer guidelines for admissions or readmissions generally fared better. Those that did not wait for PPE supplies to come from public health authorities, or for directives from CDC, and that proactively sought information, were able to move more quickly to respond. They also freely shared information received from every reliable source, including their trade associations. Policies and procedure were rewritten, supplies were pooled, and rapid adaptation of technology was used to meet new needs. It is worth noting that studies on crisis response indicate that 80 percent of us are frozen by crises, waiting for direction; 10% respond in disruptive, unhelpful ways; and 10% are proactive and take charge.

Senior care settings, like many, were forced to make do with what they had. Assisted living centers and nursing homes were at the bottom of the list of personal protective equipment but many were creative. While the pandemic was the leading story of the daily news, only months after it became evident how adversely senior care settings were affected was attention paid to the alarming death rates in those settings. Testing for residents and staff lagged even after the high death rate became apparent. So, among the lessons being reinforced was that the resource allocation between long-term care and acute care continued to be out of balance. And many were unnerved—but not shocked—by some of the cavalier behavior of those who felt they were not at risk.

Yet with challenges come opportunities. The availability and use of virtual communication technologies for residents to create easier, more regular family connections were normalized. Recognizing and responding to the deadly impact of isolation made the creation of new types of social recreational opportunities for those living in congregate settings and those living alone more common and a higher priority. And nurses outside of hospitals working in community and long-term care settings confidently demonstrated their expertise.

Leadership has come from residents as well: Consider the hundreds of older adults in congregate settings across the country crafting thousands of masks for their communities. Hold that image—untiring fingers at work, stitching together society's shared future. Perhaps it is so, that once a leader, always a leader.

A welcome, if peculiar, quality of the senior living industry historically has been the willingness of its leaders to share quite valuable information about operations with their competitors. What I describe as "collaboration not competition" has served the industry well, enabling an overall rise in the quality of care it delivers to its customer residents and their families. I hope that the spirit of transparency and collaboration becomes even greater, as the challenges—and opportunities—ahead will also become greater.

The aging of the Baby Boom population presents a huge opportunity for the industry. We know the pertinent data: Boomers comprise under 25% of the U.S. population but have more than 50% of its wealth; and within 10 years, all Boomers will be over 65. That means the industry will see an explosive growth in the need for its products.

However, if we don't figure out how to meet the demands of the middle/low middle-income market, which is currently not served either by private pay or subsidized nonprofits, we are going to be sitting with a bunch of empty properties while also seeing a tragic number of unserved older adults facing a lack of services or even homelessness. We must start thinking about how to modify the existing housing stock, develop new business models, and innovate future building because longer lifespans are going to strain or obliterate people's resources.

In terms of older people having resources, we are at the "best" moment right now that we're ever going to see. Key factors like pensions, savings, and family financial support are stronger for the current generation of seniors than the coming one. We're about to embark on a tumultuous future for the housing of older adults, unless we as a society and an industry *do something*.

There are promising possibilities on how to address this issue:

- More intergenerational housing, to create mutual support systems.
- More embedded technology to help people communicate and coordinate resources and services.
- More coordination with the health system.

All of these will be needed to provide a decent lifestyle for those entering later life without sufficient resources. How can we provide a life of dignity and respect, and at least some comfort, to millions of non-rich and non-poor Boomers—that is going to be a fundamental challenge for the next generation of leaders.

Moreover, it will take genius from many disciplines to devise housing and services for so many boomers with limited resources. We will need engineers to create integrated usable, friendly, inexpensive technology to help support people. We will need those in public health and public policy to coordinate and maximize resources. We will need financial wizards to develop new models of funding. We will need interdisciplinary perspectives, from social workers, community health advocates, gerontologists, and even artists. All are unique, and all are valuable. I urge new leaders to seek cross-fertilization in their thinking and in their actions as they try to address this issue.

None of this will be easy. Managing change and innovating is hard enough, but it will be harder still in an industry where you have prolonged, direct contact with your consumers; an ongoing interaction, not a transaction, like "here's your cup of coffee." There are so many different potentially conflicting interests: the residents, their families, their doctors, your staff, the community, policymakers and regulators.

It is important to emphasize that senior *living* is not confined to the issue of housing and services. Our aging population is part of the fabric of the community, and we cannot afford as a society to continue to treat them as an afterthought, an inconvenience, a burden. The numbers are too great, and the impact will be too profound.

So, how best to prepare for a path to leadership in the senior living industry?

First, I will emphasize a point I made earlier: *Care*. Be passionate about your work. Be sure that you want to devote your professional life (which, if you love your work, is indistinguishable from your personal life) to helping older adults live safer, more satisfying, and dignified lives. Second, look to merge theory, policy, and practice, as I have been fortunate to do in my career. To the extent you can get alignment among them, your ideas will go further and land with greater impact. Third, I urge aspiring leaders to *join and interact*. Starting when I was a student and continuing throughout my career, I have belonged to every professional association known to gerontologists, and then some. I have served on committees, boards, and working groups, and still do. Joining *inevitably* leads to relationships, collaborations, and learning. The connectivity that comes from joining will be helpful to you in ways you never expect. Finally, go out of your way to meet new people, read new topics, and do new stuff. Never be complacent with where you are. You grow, both personally and as a leader, by exposing yourself to what is *new*.

I believe that everybody has the ability to lead *something*, and to make an impact and achieve things that are important to them. You can be a great and impactful leader of a large company, an annual community fundraiser, or of a small team at your job. Leadership consists ultimately of the actions that you take to have an impact on people's lives. Leadership is achieved by saying this is important to me, and I want to be a part of something that's bigger than myself.

END

Dr. Keren Brown Wilson is the President and CEO of the Jessie F. Richardson Foundation. After her career in the development of assisted living Wilson developed methods of compassionate support for vulnerable elders in Nicaragua and most recently in rural Oregon, while helping local communities cultivate their own solutions.

Chapter 4
Ending Senior Housing as We Know It

Bill Thomas

My path to leadership began with a classic but underappreciated first step: student government. Elected president of my college student association, I gained both considerable influence on campus life and rich leadership experience, working side-by-side with the school's executives. This also taught me a profound lesson: the levers of power and influence are right in front of us all the time, but few people choose to grab hold of them. Aspiring leaders would be well-advised to determine where the power vacuums in your organization or industry are located that need to be filled.

Considering that I had this opportunity as a *student*, there is another important lesson to be gleaned: credentials, such as degrees, are neither necessary nor sufficient prerequisites for leadership. Indeed, credentials often merely help people assimilate within existing bureaucracies and are peripheral to the act of leadership. Most of us respond much more powerfully to model leadership than to technocratic leadership. That is, given the choice between an impassioned lay person who leads by excellence and example, and a highly credentialed expert with no drive or vision, people will follow the former, regardless of credentials.

This occurs frequently in the field of senior services, and it can be maddening to those who do not realize how leadership really works. Too often you will see hospital medical directors barricaded in their executive suites doing paperwork who just cannot understand why the overnight nurses have more influence on how the floor functions than they do. Too many people confuse titles and credentials with actual leadership. In this representative case, the nurses have taken hold of the levers of power.

This perspective affected not only my approach to leadership but also my career path. After graduating from Harvard Medical School, I initially practiced emergency medicine, which was exciting and gratifying. But through a part-time job at a

B. Thomas (✉)
Changing Aging, Ithaca, NY, USA
e-mail: wthomas@changingaging.org

M. Lifschultz, E. L. Schneider (eds.), *Paths to Leadership in the Senior Living Industry*, International Perspectives on Aging 29,
https://doi.org/10.1007/978-3-030-53966-5_4

nursing home, which I took just to earn extra money, I fell in love with the field of aging because of the people I served. I also observed that the field was handicapped by a lack of leadership, which I regarded as both a systemic problem and a great opportunity. Too many policies and practices were in place merely because they always had been there. I sensed and observed that the field needed reform and rejuvenation, so I changed my medical career to become one of America's few geriatricians at that time.

I was fortunate to have been mentored by Dr. T. Franklin Williams, who is considered one of the "founding fathers" of geriatric medicine. He was the director of the National Institute on Aging and mentored countless geriatricians. He captured the essence of senior care when he wrote, "The work of geriatrics is to rid ourselves, our society, and even our language, of the numerous negative terms, stereotypes, and myths concerning aging. The children, friends, and healthcare providers of older people need to give their symptoms the same respect and attention as those of younger people." Dr. Williams always focused on the person in front of him, whether it was his patient or his student. He helped reinforce my belief in myself and had a life-changing impact on me.

Early in my career, I visited a patient who was living in a nursing home. Her chart stated, "arm rash." This required a standard protocol: friendly introductions, examination, diagnosis, and prescription for treatment. About to excuse myself once I finished the exam, she gripped my arm and pulled me down so that we were face-to-face. Admittedly, it was the first time during that visit that I had *really* looked at her. She then said to me, "I'm *so* lonely." I was frozen by this moving, intimate, and sad confession. My medical training in so-called bedside manner provided no clues on how to respond properly. The moment passed. But I was haunted by the fact that, while I was treating a rash, this woman was dying of loneliness.

Out of this one bracing experience came my inspiration for the Eden Alternative Model, a philosophy for changing the culture of nursing homes, specifically to eliminate what we called the "three plagues": loneliness, helplessness, and boredom. The philosophy has spread around the world through our international nonprofit organization (http://www.edenalt.org/), and it is used in all types of care settings for the aging, not just nursing homes. The philosophy asserts that no matter how old we are or what challenges we live with, meaning is found in life through continued growth. The Eden Model helps communities end suffering and promotes purpose for aging citizens *and* their caregivers. It came out of a simple call for relief from a person in pain, and from my focus on the person in front of me, à la Dr. Williams.

Unfortunately, the field still suffers from benumbed routines and practices that perpetuate existing models of care and the *status quo*; its incumbent leadership rarely offers something new. Yet, for the next generation of leadership this is good news: the opportunity is wide open to change and improve this world. Now, it would be naive to ignore *why* the rut of creativity in senior care exists. Much of it is explained by ingrained interests that are hard to dig out from, such as the emphasis on care facilities as real estate and the entrenched politics of public policymaking. But we must pause to consider, if telecommunications had not been disrupted, we might still be sending telegrams instead of text messages. This deficit of bold, innovative, visionary leadership has done more to define the field of senior living than anything else. To aspiring leaders, I advise that you seize opportunities that exist and fill this void. Your journey should begin with an open mind and an open heart.

Another way to seize leadership in the field is to identify a seemingly intractable problem and attempt to solve it in a novel way. The horrible ignorance, fear, and prejudice around dementia until recently was obvious, yet little was being done about it. I decided to educate people about aging and living with dementia in an engaging and perhaps even entertaining way. I set out to accomplish this through performance with a theater troupe, mainly because I am a musician. This is not exactly a conventional approach; indeed, it may sound pretty kooky. We developed The Changing Aging Tour (https://changingaging.org/tour/), which features theatrical storytelling to serve this important societal objective. We blend medical science, storytelling, and live music in order to shatter our culture's damaging myths about aging and dementia. We have performed for audiences for over 5 years, in more than 125 cities, and we are still on the road. I hope that we have changed more than just a few hearts and minds.

If you have a bold and original idea that survives contact with reality—and indeed many do not—consider it an *obligation* to bring it to life. Ignore the skeptics and lead the way! Over the years, I have heard too many rueful accounts of people who had ideas just like mine but who did not pursue them because challenging the *status quo* can be hard, discouraging work. Our Green House Project (https://www.thegreenhouseproject.org/) is another example from my career. We were not the first people in the field who thought of replacing nursing homes with small, home-like environments where people can enjoy rich, full lives. Yet until we started this organization, no one ever did it. Now there are hundreds of Green House homes all over the U.S., and the movement to kill off the dreaded nursing home is strong and getting stronger.

It is critical for aspiring leaders to understand that great ideas, or even bold actions, do not *by themselves* make you a leader, or at least not necessarily an effective one. There are three keys to leadership: clarity of vision; clarity of thought; and clarity of expression. When leaders do not rise to their full potential, deficits in these areas are usually implicated.

First, to lead one must be crystal clear about your destination; that is, at what outcome do you want to arrive? Without that clarity of vision, one may take a path that will lead yourself and others to nowhere in particular.

Second, to lead one must conceive a roadmap for a path that others are not taking. Without that clarity of thought, you will end up in the same "traffic jam" of sameness and mediocrity as everyone else. Applied to aging services, this is, for example, how we got the institutionalized environments of nursing homes.

And third, to lead one must be able to clearly communicate that destination and those directions. Without that clarity of expression, everyone you lead will just get lost.

These attributes are a mix of innate and learned abilities; some people naturally have more talent in one area or another, though no matter where you are on the continuum you can improve. It takes a lot of practice, and that practice is a continuous process.

Often after I give talks, people tell me that my ideas are so simple, so easy to understand. *That* is clarity, and achieving it is long, hard work. Ironically, the better you are at it, the simpler it appears. Recognizing the difficulty and value of clarity of expression, the mathematician Pascal wittily observed, presumably after going on and on, "I would have written a shorter letter, but I didn't have the time."

The senior living industry at this moment—a "moment" that will last at least a generation—is conducive to new leadership primarily because, as currently constituted, it is a historical artifact. It is a social construct that emerged only recently, after World War II. Senior living personifies America's foray into age segregation on a mass scale, which resulted in concentrated islands of frailty and weakness that the industry perversely regards as a solution, a success. There have been many variations on a theme, but fundamentally the concept has not evolved. My strongly held personal view is, neither the current concept, nor the term "senior living" itself, will survive. To the extent senior living as we know it does evolve, it will do so first into obsolescence, then extinction.

One could characterize my point of view as both a hope and a belief that today's senior living industry will be replaced by an "independence industry." To envision this concept, think of a camera lens with three filters laid over one another, which, depending on how you manipulate them, will slightly alter the image you get. Those three lenses are architecture, culture, and technology. These disciplines will interplay and result in housing that supports independent aging, including those who are frail or cognitively impaired. We will create housing that accommodates our changing personal needs while substantially replicating the way we currently live.

My vision is based in part on the cultural aspirations of my mentor, Dr. Williams, that are so profoundly important to me: to rid our culture of ageist bias and stereotypes. The way we must do that is by telling a new story about aging as a source of well-being, strength, purpose, and belonging. Despite the fact that you may find the occasional article that extols the benefits of aging, the focus of geriatrics and gerontology is on weakness, frailty, disconnection, and social isolation. The dominant narrative is negative. We must reimagine the culture to reflect the *reality* of the beauty of aging.

To date, the application of technology to the field of aging has been saddening and uninspiring. Technology has primarily been deployed in the management of dependence, instead of protecting and extending independence. Now, I *am* optimistic that this will change. However, that change is unlikely to come from within the industry.

Recently, the American Institute of Architects honored me with their Decade Award for having the most influence on design and architecture in the past 10 years in the field of aging. But while that was nice, it is also disappointing, because the award was for recognition of the Green House model, which was launched back in *2002*! There should have been ten better ideas since then. Or a hundred. The Green House should have just been swallowed up by a wave of cool new architectures, technologies, cultures, and systems. While there is a timelessness to our design, which we based on the Danish concept of *hygge*, or cozy charm, the field is ready for a flood of creativity—a flood that will wash away the current landscape and open up the field to creative destruction, reimagining, and rebuilding. The flood is the only metaphor in that vision; the rest is literal.

The flood is an apt metaphor, though I prefer to call the changes coming to senior care and living as the Age Boom. There is no doubt that the aging population is swelling and along with this demographic transformation will come changing needs. I believe that the Baby Boomers will come to view today's senior living and long-term care the way people view the life vest on an airplane: you're glad to know it is there but pray you never have to use it. It will not be a sought-after lifestyle choice. That rejection will, thankfully in my view, hasten big change, which I predict will originate from the dialogue between Boomers who do choose senior living and the industry as it finds a new "identity." That conversation will go from the prevailing top-down focus on "what they need" to the consumer-driven "what we want." To succeed, existing operators will have to understand and execute that pivot, but like a large ship, that will be a hard turn to make fast.

But I have faith that change is eminently possible. Indeed, that is why we just recently launch a new venture aimed largely at what is called the "missing middle," those seniors who cannot afford private pay senior living but also do not qualify for housing subsidized by Medicaid and other government programs. Our organization has developed Minka, which is a Japanese concept that translates as "houses of the people." Minka is a builder of affordable homes and communities using an innovative, adaptive modular building system, and 3-D printing to make housing that is designed for its dwellers (https://myminka.com/).

Minka's creation story started with architecture and design. Since the end of World War II, the average size of an American house has doubled, yet the average number of people living in an American house has halved, so on a *per capita* basis, four times as much space is being used for occupants of houses today. Minka focuses on smaller, more compact independence-oriented designs that can be acquired and lived in for less money, and therefore can address a market need that is going largely unmet. Minka combines culture, architecture, and technology to protect and extend the independence of its residents. By placing Minka homes in dense clusters, we will help foster communities that create a culture of reciprocity and mutual reliance—and *that* is far better than being stuck in your big house at the end of a cul-de-sac where you don't see anyone all day, like so many of today's seniors, especially in the sprawling suburbs. We have a very clear perspective on how those houses can be used to create communities. This vision is brought to life in my novel, "In the Arms of Elders." We are hacking the homebuilding industry because large, underused houses that cost a lot of money are bad for older people; indeed, as a geriatrician, I can say that such houses actually *kill* many people.

Needless to say, senior living is not an industry known for its creativity. To be fair, this is due partially to the phenomenon that success by its very nature tends to degrade and diminish creativity, and this industry did find formulas for success, and stuck with them. The paradox is, success and creativity often work inversely to each other, and ultimately that guarantees failure. Yes, we all can name individual artists who remained creative and successful throughout their working lives, but it is very difficult for organizations to do this. This makes intuitive sense: when the affairs of a business are going well, organizational autopilot kicks in and people feel that maintaining the status quo is their best career-advancing move, rather than thinking up disruptive new ideas. Under these circumstances, if you want creativity you will need creativity sessions led by a creativity consultant who can provide corporate shock therapy to try to get the creative blood flowing again. Doesn't that sound awful?

We focus on the elusive line that separates truly impossible ideas and those ideas that are only *mostly* impossible. I must cite the wonderful film "Princess Bride," where there is a difference between dead and mostly dead. *Mostly* impossible ideas are the Holy Grail, but they are surrounded by ideas that are totally impossible. Culling these apart is the practice of the creative act, understanding which side of the line you are actually on. The mostly impossible idea—*that* is where you want to devote your time and energy.

Surgery may be an apt metaphor for leadership of a creative organization. People may assume that surgery is about not making mistakes, but in practice it is about knowing how to *fix* them. If you nick an artery and do not know how to repair it, a patient dies; but if you can react quickly to repair it, the surgery proceeds and the patient lives. Moreover, many surgical procedures are inherently risky, but the reward for success is so great, you do everything you can to manage the risk *and* proceed. So, in an organization that wishes to be, or even depends on being, creative, you must cultivate the ability to fix mistakes. Because creativity is inherently risky; by definition it involves venturing into the unknown, which unavoidably stresses an organization. However, creativity can improve the health of an organization and extend its life. Many would-be leaders act with the belief that you get just one shot, that if you make a single mistake, you lose everything. This can become a self-fulfilling prophecy, as it creates a cognitive distortion that causes such a leader to either exaggerate or minimize risks, thereby avoiding creative advancement.

Creative risk-taking has certainly been the hallmark of my career. My work has been characterized by taking on new endeavors in seven- to ten-year cycles, including practicing medicine, Eden, Green House, Changing Aging, and now Minka. With Minka, we are in the early part of the cycle and especially in this phase, I will get great inspiration to grow as a leader because so much is unknown. We do not know exactly what we're doing, though we do know what we want to achieve. Trying to figure it all out is a constant challenge. You are learning and doing new things *every day*. The deck is stacked against us and the stakes are high. If that doesn't inspire you to continue developing as a leader, what will? As T.S. Eliot said, "Only those who risk going too far can possibly find out how far one can go."

As I write, we are approximately 6 months into the COVID-19 pandemic, and a question that is frequently asked is, how will the pandemic change the senior living industry. I believe that it will *change* nothing. If that seems an unusual opinion, I have two reasons for holding it, one derived from optimism and the other drenched in pessimism. So, let's get the bad news out of the way. COVID-19 changes nothing because we have mountains of evidence generated over decades by thousands of researchers showing that preventable infections are killing elders in senior living environments. Rates of infection transmission could have been slashed long ago and hundreds of thousands of lives could have been saved from hand washing, basic hygienic practices, and standard infection control techniques. Instead of making progress in this area, we find that Infection Control F-tags continue to be commonly cited by surveyors. COVID-19 won't change this grim reality because we, as a society, expect people living with frailty to get sick and die. Those ageist expectations sap energy from reform and lead to a collective shrugging of our shoulders.

On the brighter side, we can see COVID-19 accelerating trends that are already well-established. The trend toward home and community-based services will accelerate. The trend toward greater use of technology will accelerate. The trend toward population-based systems and models will accelerate. COVID-19 will not change our direction, it will just move us toward this much brighter future at a brisker pace.

It is hard for me to think of COVID-19 in such dispassionate terms, however, because the loss of life is so immense. It seems to me that a tragedy this great must lead us to something better. Shakespeare, for his part, understood that all real change is founded on tragedy, and he exclaimed, "Blow winds and crack your cheeks!" The tragedy has come, and change may follow, if, this time, it springs from more than good intentions.

With or without the pandemic, we are at a decisive moment not just for senior care and living but for society at large. The aging industry's understanding of what we *need* must be entirely transformed into an understanding of what we *want.* Like being on a teeter totter, those of us in the field of aging must decide, which side are you on? With the explosive growth of the aging population worldwide, challenges and opportunities are going to come at us with unfathomable speed. Why not be as bold and ambitious as one can be? Why not invent new architectures, new cultures, new technologies? Why not re-mix human wealth, social capital, financial capital in new ways that support people's quest for autonomy, dignity, and independence? Be generous with yourself and bring that generosity to all your endeavors.

Others may say, quite legitimately, just get a good job, secure a future path, and enjoy your time working with elders. But as we head rapidly into this new world, I am urging the leaders of today and tomorrow to cast aside the leadership of yesterday.

END

Dr. Bill Thomas Named by the Wall Street Journal as one of the top ten Americans shaping the future of aging, Dr. Bill Thomas is a Harvard-trained physician, author, performer and serial entrepreneur whose work has focused on improving the health and well-being of older people. In the early 1990's he co-founded the Eden Alternative with his wife Jude Meyers Thomas. This approach brought plants, animals, children, and news ways of living and working into sterile nursing homes and soon spread around the world. In the 2000's he developed the Green House™ model that enabled older people living with frailty to become part of a community, living in a household of ten to twelve people. Dr. Thomas is winner of the American Institute of Architects' Decade Award for greatest influence over design.

He is now focused on helping people of all ages and abilities live in the place and manner of their choosing. The simple design of Minka is based on research into human complexity and decades of input from some of our culture's most vulnerable and valuable people. Dr. Thomas has distilled these insights into the Minka Building System™. The resulting homes and communities are optimized to help people move, eat, sleep and heal. Working with the University of Southern Indiana, Dr. Thomas and his team are developing a community-building model, **MAGIC** (**M**ulti-**A**bility, multi-**G**enerational, **I**nclusive **C**ommunities).

Chapter 5
Applied Evangelism in Senior Living

John Erickson

After seven years of college and graduate school in the Catholic seminary, I began my career as a systems analyst for mainframe computers. It was well before the modern age of personal computers. Since the machines each cost millions of dollars and required 100 tons of air conditioning, I thought that growth in this industry would necessarily be limited. So, I left the field to put together syndicated real estate investments, which ultimately led me to acquire thousands of housing units in Florida.

The residents were primarily retirees in their 60s. It was a great learning experience, because I was involved in almost every aspect, from operations to finance to marketing. Being hands-on also enabled me to become acquainted with many of my customers, which was a very gratifying part of the work. This led me to the insight that the *real* opportunity was not in Florida, because for every one older person that moves to Florida, twenty others stay home. That meant that the market was 20 times larger elsewhere. This led me to consider developing a senior housing concept for the larger cities up north.

Moreover, I concluded that a housing concept for older people ought to be aimed at 75-year-olds. My theory was that those in their 60s who stayed in their homes did so for compelling reasons, such as family, church, and community. Their minds would not be easily changed to move into dedicated senior living. However, at 75, many would be quite receptive to a new lifestyle, one that addressed the burdens of housing upkeep and health issues, while featuring the appeal of vibrant social living. We tested this theory by re-developing an abandoned 120-acre college on the edge of Baltimore into the first Erickson mega-retirement community.

There were many aspects of the concept that were novel, especially the financial terms. The prevailing formula for continuing care retirement communities (CCRCs)

J. Erickson (✉)
Erickson Living, Catonsville, MD, USA
e-mail: john.erickson@erickson.com

M. Lifschultz, E. L. Schneider (eds.), *Paths to Leadership in the Senior Living Industry*, International Perspectives on Aging 29,
https://doi.org/10.1007/978-3-030-53966-5_5

had been developed in the 1950s by the Quakers; it required a substantial down payment, also known as buying in, and it was suitable only for those of high means. For my new company, we devised a revolutionary financing model, one based on a refundable sum, that not only put our housing within reach of middle-income folks, but also included all the basic services, like meals, utilities, taxes, and transportation. Because of the aggregated environment, or economies of scale, we could create a support system of services that would be far cheaper than the costs on an individual basis. This provided us access to an enormous potential market. By innovating the existing economics, we moved the previously exclusive CCRC concept into middle-income America. That was a huge step for the industry.

However, bold innovation often meets resistance from incumbent institutions, and in our case, this came from the banks, which deemed our financial structure to be unconventional and potentially nonviable. It took much work and some good fortune, but ultimately, we were able to secure financing support from the GEICO Insurance Company, and we were "off to the races."

Since the middle-income market was unserved and we were introducing this new financial model and broader service features, demand for our product grew rapidly. The prevailing theory of the industry had been, do not build a senior living community containing more than 200 units, because putting together too many older people at the same time will inevitably fail, acting like an atomic reactor out of control, yielding an unmanageable community where no one would ever want to live. In practice, we proved that this old theory was completely wrong. The more people that we put together, the more robust the community became. Activity simply increased exponentially: more clubs were founded, more bridge games occurred, more volunteerism resulted, and demand rose for social destinations like dining rooms and libraries. The critical mass created more vibrancy. With 1600 independent apartments and 400 health care beds, the community was an outstanding success and established new horizons for the senior housing industry.

Despite the successes of our communities over thirty years, the challenge of maintaining access to the capital markets still remained a daily challenge. Indeed, this became virtually a full-time job. The operations—dealing with residents and managing services—was great fun, but soon after we started growing, I effectively moved into a corner office and spent most of my time in finance, while delegating the operating side to our management group.

Fortunately, we were never short of talented people who were passionately interested in our work, so I was able to focus on my primary work while delegating the rest to excellent executives and managers. Our company was truly mission-based, in that we were providing senior housing for middle America, which previously had been ignored. Moreover, our overall objective was *excellence*, to do housing for seniors better than anybody else had ever done it. We became the star of the CCRC industry. At industry conferences, everybody wanted to know, what is the Erickson company doing now, what are they doing next? By being an industry leader, we were able to attract the industry's best talent. When we could not find those with experience because it was a fledgling industry, we were able to attract talented people who could be educated and re-trained from other industry sectors. The key to our success was that we were a mission-driven company, and people liked being a part of something bigger than themselves.

A great feature of our industry in the early days was that we pioneers all shared our knowledge, experience, and wisdom. I did so enthusiastically. Today, one might call it "open source." Whether it was sharing best practices at conferences or arranging facility tours for would-be competitors, we banded together for the good of the industry and especially for the good of the residents we were serving. This did require a confidence of leadership, because you were revealing your trade secrets, but the industry as a whole certainly benefited from it. It was a form of "coopetition." Perhaps the most important by-product of this was that it helped defeat the real enemy of the industry, which was a shortage of capital. By coming together, we demonstrated that senior housing was a qualified financial niche, worthy of lending and investment from the banking community, REIT's, bonding authorities, insurance investors, etc. Our efforts together grew those financial markets, which in turn helped enable the industry to grow.

Of course, with a growing market comes challenges and opportunities, and once our company was large, the stakes got ever higher. For example, by staying in touch with so many residents, I observed up close that health care was a huge issue. Imagine an 80-year-old with various medical conditions seeing five or six different doctors, none of whom were coordinating care with one another. Appointments took weeks to arrange. The outcomes were often dismal. We decided to develop our own health care system, and it was one of the most complicated efforts we ever took on. However, as a leader with not only people's *quality of life* but also their actual *lives* in our hands, I was not going to shrink from the challenge.

We knew that we needed a full-time physician on our large Baltimore campus. After attempting unsuccessfully to get a local hospital to detail one of theirs to our community, we simply brought on a full-time geriatric physician, which quickly led to hiring several more, because the response and need for primary care was overwhelming: the residents were moving their care to the on-site physicians, and their outcomes improved dramatically. To "justify" the expense of having our own doctors, we did rigorous studies, comparing the care and outcomes of our residents with others who were managing their care through Medicare from their homes. It became clear that the coordination, the ability to get a timely appointment to see a physician, and our pioneering use of electronic medical records, resulted in superior healthcare. After years of relentlessly lobbying in Washington, I was able to get the federal government to authorize a demonstration program that effectively gave us our own Medicare insurance program. I was, again, pushing the boundaries within our industry, challenging long-held assumptions and practices. The benefits of these efforts were confirmed by studies that proved that our approach to care for seniors was vastly superior to conventional practices, particularly when one looked at indicators such as length of hospital stays and re-admission rates

Today, the issues of connectivity, isolation, and loneliness among the aging is getting intense—and deserved—attention. We identified this early on, noting the correlation between connectivity and people *wanting* to be healthy. For example, if folks are missing their bridge game due to illness, they are going to try harder to stay healthy. But if they are at home, alone, and the only activity they have is watching television, they will be less motivated to tend to their health. Therefore, we focused on creating the best possible environments for socialization, exercise, and other congregate activities, and combined this with our on-site medicine and better pharmaceutical management. Thus, we achieved great results for our residents.

Achieving great results was not accidental. It was the intended effect of talented people coming together to execute well-conceived strategies. I believe that the most important characteristic of leadership is a well-defined vision. It was my passionate commitment to a clear vision that encouraged talented and committed people to come work alongside me. Without them, I could not have achieved what we did. As a leader, you must be able to get others to buy into your vision, internalize it, and execute it with you seamlessly so that you are not trying to do everything yourself. This also requires strength of personality. A leader with a great vision but a weak persona will have trouble getting followers to get on their path.

Specific to senior housing, the most important trait for a leader is that you must care a lot about older adults. It must be meaningful to you to be in their lives. Personally, that provided me more value and satisfaction than any other aspect of the work. It is a privilege to hear these incredible life stories every day, about what people had achieved in their lives; their trials, tribulations, and triumphs. Few people in our business do it just for the bucks.

Because you are dealing with people's *lives* in this industry, it is especially important for leaders to be willing to cut losses when it comes to personnel. Bad work, bad work habits, bad attitude—they are not only contagious to other employees, but they also can be a threat to your residents' well-being and happiness. Considering that even when you make excellent hiring decisions you will be wrong approximately twenty-five percent of the time, leaders must be decisive in firing. If your own people are holding you back from achieving your vision, then really you are not leading.

To be effective, a leader must find managers who believe in the mission; indeed, that encapsulates perhaps the primary requirement of both leaders and managers. What is often lost on leaders is that the roster of managers directly reflect upon the effectiveness of a leader. This means that leaders must be exceptionally discerning at picking managers well. For example, there are two, all-too-common managerial types to look out for: one, those who are quite good at working with their boss, but very poor at working with everybody else, particularly those they manage; and those that are threatened by having strong employees underneath them. Both of these traits demonstrate a weakness of leadership, as good leadership requires having the confidence to build around you the strongest team possible. Good leadership means giving people enough runway that they can have the most impact and create the most value within their potential. Leaders must trust and believe in those who work for them in order to grant those people that autonomy.

Attracting great people is difficult, but retaining them, especially in an industry like ours that demands so much of its professionals, is even harder. Above all else, you must make them feel good about where they work. This may be elementary, but it is worth emphasizing: You must always pay people in the upper half, if not the upper quartile, of the industry pay scale. This is a necessary but not sufficient condition, however. Even more important is enabling your people to find meaning in their work. If people are working in senior living only for the paycheck, you have either the wrong people or the wrong environment, and you will not enjoy strong retention. But if they are there because of your mission, then they are going to be much more inclined to stay. And that will cut across the entire organization, from frontline workers to executives. For example, if one treat nurse's aides as if their job is merely to bathe residents, that could be quite demeaning. But if one regards their work as giving residents comfort and self-respect, then together we have a meaningful mission.

On a tactical level, I believe deeply in connectivity programs, through which you create strong bonds between your people and your mission. It is contagious—in a good way—when you recognize employees with awards for their manifest commitment to your values. For example, we had a robust college scholarship program for high school students working in service jobs that benefitted these employees, the residents, our organization, and the greater community. Such conscious efforts will keep retention high, even in our increasingly mobile economy.

Overall, the advent of the Baby Boomer generation should be expected to bring value to the industry through the social side of senior housing. Many will choose to reject the isolation of staying in their own homes and instead seek out social environments for healthy aging. Thus, there are good years ahead for this industry, particularly for those organizations that provide full-service housing, like Erickson.

Opportunities for the American senior living industry can also be found abroad. In China, for example, the market may not yet be ready for our approach to senior living, but the country has far more older people than we do, and as it develops, China's educated population is going to demand more dignity and service for seniors than is currently provided. That market will increasingly accommodate American participation, as our concept of having a critical mass of services is ultimately the only way to provide good care for a community of older people. Our know-how should be welcomed, though that market may not ripen for another ten years. However, it is not too early for entrepreneurial leaders to be positioning themselves to pursue it.

In Northern Europe, historically they have relied on the home care model, and that probably will not change even though they too have rapidly growing aging populations. Their more homogenous societies, where most people live within a much narrower range of poverty and wealth, means it is unlikely that American concepts will take hold, since they result in part from the inequities of our distribution system.

We are still at the infant stage of developing senior services, in that leisure retirement is a relatively new notion; until the 1950's, most of us died at our "workstation." Now we have an unprecedented span of discretionary time, and we have not entirely solved for ourselves how and where we will spend it. Less than five percent of people can now make a choice that works for their living needs, which means that

95 percent are not going to get a decent option unless and until the industry adapts to them.

Our industry suffers from a largely negative association or brand that is a being perpetuated by outdated products and services. For example, the physical plants of many seniors' communities are 30 years old. And the perception of "warehousing old people" has to be educated out of the public's perception.

One way that industry players can stay ahead of the change curve is by fomenting creative thinking. It must be a leader's priority to pursue innovation and invite it from residents, from staff, and from outside intelligence. This will, inevitably, affect culture and product in positive ways. Too many companies do the opposite, erecting barriers to innovation, protecting the status quo, and retaining all the decision-making to the corporate office, which is removed from where the actual residents are living. That makes no sense in senior living. You want to push down to the ground level as much problem-solving and idea-generation as you can, and the result is greater responsiveness and more vibrancy

Efforts to innovate go directly to mission, which is where a senior living organization's energy should mostly be devoted. By constantly talking up and acting on your mission—ours was to be the best in the senior living market—every day, you can advance toward that goal. By setting the mission to be foremost, you also engage your whole team to contribute to how vision and values develop. By doing this, you are leading at the top, but you also are drawing upon the resources below. A too-common failure of leaders is the belief that culture is handed down, but that is wrong; it is built up.

As a former seminarian, I like to tell the story of Moses moving the Israelites out of Egypt. Once in the desert there was nothing but complaining. Moses got fed up and told God. The Lord did not want to end the exodus there, so he empowered 70 of the elders to be as strong as Moses was with prophecy and leadership. Then Moses told God, you took away my power; that is not what I wanted. And God said, that may be, but I am leading *my* people. Moses had lost sight of the mission by focusing on his power. But when the mission was shared below, the scene was set for success.

A question that leaders must ask themselves is, where does my sense of mission come from? Such self-examination ought to reveal your mission's authenticity. I was one of fourteen children. When I was 12 years old, I met a couple at church who had no children but had a beautiful home and a lot of love to give. We took to one another and they effectively adopted me. Ultimately, both suffered from dementia, and I helped care for them. However, I had no legal rights to their welfare, and after my surrogate dad passed, my surrogate mom's family took over her care and, tragically, moved her into a nursing home to wither away. There was nothing I could do but just watch it happen, and it was a terrible ending. It caused me to appreciate that aging is quite a complicated matter.

As a result, in my work with older adults I was always trying to find better ways for people to age. The two unfortunate experiences I had with my surrogate parents was very much on my mind when I began working with those retirees in Florida, and I just knew that I had been called to help people age with dignity. If we cared about older people, the way God would like them to be cared for, the Mission becomes obvious. This realization became the guiding principle in my life and was an everyday part of the Erickson culture and the foundation of our Mission, Vision, and Values.

I believe mission comes from one's internal sense of being, but sometimes we also find inspiration coming from a source outside ourselves. And if one is fortunate, along the way in our leadership development we find an inspiring figure from whom we learn great lessons. For me, that person has been Dr. Freeman Hrabowski, the African American educator, author, mathematician, and president of University of Maryland-Baltimore County, which Dr. Hrabowski turned from a local commuter school into one of the nation's most respected public universities. His university generates more students of color into the top MD/PHD programs of American Universities than any school in the nation. As a 12-year-old, he was among the *leaders*, along with Dr. Martin Luther King, Jr., of a great civil rights march on the courthouse of Birmingham, Alabama. He was met there by the infamous racist sheriff, Bull Connor, who spit on him and arrested him on national television. For marching, he was expelled from grade school. But he went on to graduate from high school at 14, from college at 17, and to earn his PhD in mathematics at 21. Some of the qualities one must use to describe Dr. Hrabowski are bold, tenacious, and innovative. He is a true leader.

In reflecting on one's own mission and inspiration, one is likely to consider their own style of leadership. My self-perspective is that I practiced what could be called "entrepreneurial leadership," whose distinctive feature is an unusually high tolerance for risk, one that is higher than is sometimes healthy. This style can lead to the mistake of operating with imbalances between demonstrated risk and reward potential. You might call this "pushing the envelope." When you have self-identified as a risk-taker, a corrective measure is to use your management team and/or your board as a governor and a guiding influence, while still retaining to yourself the decision-making. It takes keen self-awareness and strong self-confidence to be such a leader.

A potential outgrowth of a high-risk/high reward leadership style is a constant-growth environment where mistakes inevitably occur, in part because it can be hard to manage evenly when you are running a marathon at a sprinter's pace and not everyone around you can maintain that speed over that distance. In that vortex of growth, as a leader you may expect that everyone can keep up, but then as a result, you get some people burning out, others doing whatever they can to keep pace, and still others entirely losing sight of the finish line. It is up to you to keep the pack running together, or else the organization will achieve very little except disarray.

Ultimately, leaders are born, not made. Leadership is more of an innate characteristic and talent than a developed quality. You either have leadership capacity or you don't, and one is unlikely to gain it by practice. However, if one is a natural leader, then you can get even stronger, by working with and comparing notes with peers. I encourage aspiring leaders to seek out one another and commit to working together on leadership development whether as individual coaches or in peer groups. Both of these approaches are invaluable for observing how others make decisions and for talking through your difficult decisions with those in whom you have built trust and respect.

The coronavirus creates special challenges for senior housing management. By definition our population is older and clearly in the highest risk group. The fact that this coronavirus has had such a lethal effect on this population is really not surprising and should be expected. The reality is that all contagious diseases are particularly dangerous for older adults because most seniors have multiple chronic conditions that effect the stability of their basic life systems. When these less stable life support systems come under a viral attack, it often pushes the limits of recovery and leads to the final cause of death. This is precisely the reason why we encourage every senior to take a flu shot. Unfortunately, with a novel coronavirus there was until recently no vaccine, so the choices were isolation and containment or infection and the consequences.

What is particularly unfair for the senior industry is the media attention to what is a very natural phenomenon. With our ageism bias and our culture's quest for immortality, media portrays this as a terrible event, exacerbated by human failure and error that should never occur. And yet hospitals across the nation were discharging COVID-19 patients to nursing homes, thus spreading the infections that they were trying to avoid. Why would you send a COVID-19 patient into a petri dish of older, frail, super-high risk compromised seniors? The answer is because we have a very broken health care hierarchy that gives hospitals and doctors priority based on reimbursement and has at the lowest end of the totem pole the nursing home. Sick and recovering COVID-19 patients were sent into environments that operate without doctors and mostly with minimum skilled nursing assistants, who in this instance often had no protective gear, isolations systems, or testing ability. Of course, 50% of deaths were in nursing homes, but it wasn't the fault of nursing homes—it was the poor judgment of the health care system. The flow should have been the other way: moving people out of nursing homes into acute hospitals at the earliest testing point. Hospitals will scream for reimbursement along with the doctors and give nothing to the nursing homes that carried the bulk of the healthcare burden. That story never gets written. Industry behemoths such as Johns Hopkins, Cedars-Sinai, Mass General, etc., with multi-billion-dollar balance sheets, will get full coverage, compensation and reimbursement, while nursing home operators will get regulatory fines and blame for gross incompetence, along with years of lawsuits from ambulance-chasing attorneys. I urge the leaders of today and tomorrow to help reorder the way care is delivered to our older population.

Surprisingly, though, it won't have much long-term effect on the industry overall. The reason is that there will always be older adults in need of care, and most cannot be managed in the home environment. So, life will return to the same quickly. More independent seniors will still make their best choice for higher quality living in full-service communities with services and amenities. There is an innate human quality of wanting to choose the space and environments that make us feel the best and support our social goals and desire for dignity and independence. Well managed Life Plan Communities create that desirable and necessary space and will rise from the ashes of the coronavirus with strong supporters in spite of media and society's negative views on aging.

In conclusion, I must stress that I am not through evangelizing. At 75, I'm trying to figure out just how much I can do and where I can get it done. So, expect another chapter from me in the days ahead.

END

John Erickson is the founder of Erickson Living, America's leading developer and manager of full-service campus style retirement communities, currently serving more than 28,000 residents across 25 campuses in 12 states. The Erickson mega-community model can serve over 2500 full time residents across the full continuum of care including Erickson Advantage, the only private company Advantage Plan ever approved by Medicare. He also founded national media enterprises RLTV and BOOMERS TV targeting the 50+ demographic's quest for lifetime learning and better living. A well-known philanthropist, he contributed to research in aging through the Erickson Foundation, and to improved education for underserved kids at his NorthBay Environmental Campus in Northeast Maryland. Before he retired as its Chairman and CEO in 2008, Erickson Living was named as one of FORTUNE's "100 Best Companies to Work For."

Chapter 6
Pioneering Assisted Living

Paul Klaassen

I started Sunrise Assisted Living in 1981 when I was 23 years old. Sunrise was based on a model whose advantages I knew firsthand from the Netherlands. My family is Dutch, and my two grandmothers had experienced superb long-term care in assisted living settings there. One of my grandmothers received dementia care while the other received traditional end-of-life care. They both died in these wonderful caring settings. Neither one ever went to a nursing home. Nursing homes were, and are, used much more rarely in the Netherlands than in the U.S. The excellence of the care and quality of life my grandmothers experienced left a lasting impression on me and helped to shape my career.

Not everyone opens a senior living facility soon after college graduation, but that is precisely what my wife, Terry, and I did. We opened our first Sunrise and moved into the building. We lived with the residents there for about a year and a half, and I worked at every position in that community. We marketed Sunrise as a nursing home alternative. At that time, nobody in the United States knew what "assisted living" was. It would be years before the general population recognized that we had developed a true long-term care alternative.

In the 80s, we were opening one new community every year. We built a reproduceable prototype community that would fit on a one- or two-acre site. This made our growth rapid and manageable. Soon, our growth accelerated to opening three new communities a year. Sunrise went public in the 1990's, and with our additional capital, we opened fifteen or twenty newly constructed communities annually; at our peak, we opened 28 in one year. We also moved into Canada and the United Kingdom. Overall, we built approximately 400 assisted living communities.

Our greatest challenge was surmounting a deeply ingrained public perception of care for seniors. When we started Sunrise, no one believed that long-term care and end-of-life care could occur in a setting other than a skilled nursing facility because

P. Klaassen (✉)
Sunrise Senior Living, McLean, VA, USA

M. Lifschultz, E. L. Schneider (eds.), *Paths to Leadership in the Senior Living Industry*, International Perspectives on Aging 29,
https://doi.org/10.1007/978-3-030-53966-5_6

that is the way it had been done for decades. This point of view was propagated by the medical and healthcare communities through popular culture such as television news, books, and magazines. The expectation of the public was that to get long-term care or healthcare at the end-of-life, you had to be in a place with hospital beds, fluorescent lighting, vinyl tile, and institutional food and surroundings. Sunrise simply did not conform to this stereotype.

We were combatting an ingrained societal belief that institutional healthcare was the only option for the frail elderly. But we viewed this from an entirely different perspective, and one that was less ageist. If we were caring for *young people*, they would certainly not be placed in a nursing home. We identified a double standard that many did not yet perceive—Americans were institutionalizing 84-year-olds for reasons we would never institutionalize a 4-year-old or a 44-year-old.

We *knew* that care is portable; it can be delivered in almost any setting. We knew the facts were on our side. The more you travel around the world, the more you can see that there are better ways to provide long-term care. We had confidence in our model, and in our outcomes, and that when customers experienced the difference, our less institutional model of long-term care would thrive. And it did.

We had plenty of other challenges. Raising capital and satisfying regulators were separate but related issues, and both relied on our ability to convince deeply skeptical audiences of the advantages of our model. From one state to the next, there was an arbitrariness to the rulings we faced. It was common, for example, for a state regulator to decide that a resident was too frail to be in our community, even though the family and the resident thought they were appropriately placed—and even though the care was paid for privately. Ultimately, many of the rules and regulations were changed. However, even today, this work is still not complete: We still live in a country where those with limited finances do not have access to the same settings for long-term care that private pay individuals do. Today, we have sadly a two-tiered long-term care system in the U.S., where the poor have very limited choices and experience their care in a Medicaid nursing home bed while the middle and upper economic classes have a much wider array of assisted living and home care choices.

Having started Sunrise right after college, I didn't really have the time to *find* a role model or a mentor, which would have been an helpful step for a relatively young person intent on upsetting an entire industry. Fortunately, there was already one in my life: my mom. She was the leader that I most closely modeled my career on. My mother was exceptionally hard working, highly entrepreneurial, and had tremendous composure and discipline. She founded and operated two private schools with childcare centers.

Observing my mother's work style provided invaluable lessons to me. There are close parallels between managing childcare and eldercare. Perhaps the greatest similarity is in the importance of dealing with family members, whether with the mothers of children in childcare or the children of mothers in eldercare. In both cases, the lines are somewhat blurred as to who is the decision maker and who is the consumer. It doesn't really matter since both must be satisfied and childcare and eldercare are human services businesses. My mom was very good at it, and I was fortunate to have learned so much from her.

In addition to my mother, I have observed other leaders at many other companies, and have concluded there are many different character traits that lend themselves to leadership. For the senior living field in particular, there is a leadership model that has wide application and effectiveness: servant leadership.

The principles of servant leadership, which can be found in many belief systems, such as Taoist and Christian philosophies, were brought to the modern audience by the thinker and writer Robert Greenleaf. His book from the 1970s, *Servant Leadership*, was seminal, and continues to be highly influential. The ten qualities possessed or practiced by a servant leader have been distilled as: listening, empathy, healing, awareness, persuasiveness, conceptualizing, foresight, stewardship, growth of people, and building community. This is a powerful set of principles, which many of us look for in leadership. There is a special, unique nature to the mission of championing seniors' quality of life, and the senior living field is well matched with servant leadership attributes.

Living 24 hours a day in the first Sunrise community with the residents had a formative impact on me as a leader. I not only heard the stories of residents but lived them too. I developed a deep understanding of the critical role of the family. I *saw* what was important and what wasn't important to the quality of life of a frail senior. The lessons that we gained early on from living in that community informed me for the rest of my career. Growing relatively slowly and steadily—organically—also gave me the chance to learn and apply lessons in a deliberate way.

In general, leadership requires confidence, courage, and conviction. There is little room for self-doubt and second-guessing because those qualities are corrosive and quite contagious. It is hard to follow a leader who doesn't know where they are headed.

What is different about leadership in senior care is that the *mission* is paramount; it has a moral dimension. We are not selling toothpaste or accumulating social media "likes" or merely pursuing profits. We are, typically, caring for 87-year-old frail women—mothers, grandmothers. One *must* have a lot of compassion, sincerity, sensitivity, and passion for these wonderful human beings. These may also not be bad traits for running a software company, but they are *essential* in senior care. *Caring* is both an added complexity and an added advantage that the field has in terms of attracting employees. This mission orientation translates into what many of us call the "second paycheck." The vast majority of leaders and managers who succeed in our industry do so for reasons besides just wanting and needing their salaries; they also are dedicated to serving seniors.

As a leader hiring other leaders and managers, I always looked for demonstrable evidence that a candidate fits this profile. For example, at Sunrise it was essential that applicants could show they did volunteer work, because to me that is the single best factor in identifying whether or not a person has a heart for service and a commitment to something other than just serving themselves. It would hardly matter what the volunteer work was, and it did not have to be working with seniors. But, since less than 25% Americans do almost 100% of the nation's volunteer work, it is a meaningful expression of what makes someone special and that they would fit the Sunrise culture. The fact is, working in this field matches the requirements of volunteering: it requires empathy, hard work, long hours, and a heart that loves to serve. Plus, it is 24/7, working weekends and holidays. If you only do it for the money paycheck, your career probably will not last long. By looking for people who volunteer, we found people consistent with those who stay in the field and succeed as leaders and managers.

This extends to building great teams: having a group of others-oriented individuals who each have a heart for service and a desire to work for the "second paycheck." This gives you a significant head start, because you know you are training people who share principles and values. That is the basis for an effective team. If you have some team members who are looking out for themselves, or you have four different cultures, or three different sets of principles and values, and you form a team from this, you will have chaos.

Inextricably related to teambuilding is retention, which starts up front with the selection process. Attracting successful individuals, building effective teams, and maximizing retention can be seen as parts that come together to make a healthy organization and culture.

I always need to emphasize that the work is just too hard to do if you are in it only for the monetary paycheck. Good people stay because they like to work for an organization that puts their mission front and center. At Sunrise, we start and end every meeting on our mission. We always try to bring the topic back to our mission.

It is also essential to provide a quality work environment. Sunrise has always been known for its high-quality buildings. Companies in our industry must recognize that our communities are not only the living environment for our residents, but also the work environment for our employees. Where would you rather work as a housekeeper, at a Ritz Carlton or a Motel 6? Having a pleasant atmosphere, banishing institutional lighting, providing warmth and coziness with lots of common space and elegant furnishings—these are fundamental to an environment where people will love coming to work. It is fundamental to both residents' and employees' quality of life.

Our industry does have an advantage competing for talent in today's mobile workforce, especially at the frontline level of service delivery: it has the noble mission and purpose that many jobs simply do not. Motivating people with a mission to serve coffee or stock a warehouse has got to be more difficult than motivating them with the mission to serve seniors. It is unlikely that our team members would find serving coffee at all fulfilling. As competition for workers intensifies, and the need for care workers accelerates, I believe our field should focus on attracting those who seek mission-driven work.

As we look at the future of senior living, one area where we can expect widespread change is in end-of-life care. Sunrise has been focused on it from the very beginning. When older adults come into long-term care, they and their families are aware of the implications. People are very aware that mom or dad is entering the last chapter of life. And they want the best possible experience for them.

Americans currently do not die the way they want. Most people want to die at home surrounded by family and friends; lying in their own bed, on their own pillow, looking at their familiar art on the wall. But overwhelmingly we end up dying looking up at fluorescent lights, perhaps with a tube up our nose, surrounded by strange people and sounds and whirling machines. This should not and *cannot* continue. There are some signs of improvement; but we must keep in mind that, while the assisted living movement has changed long-term care dramatically, it has taken 30 years. End-of-life care is not making nearly as much progress as I would have predicted 30 years ago. Most people still die in places and ways they don't want. There is way too much unnecessary and unwanted medical intervention at the end of life. And, perhaps most problematically, regulations and healthcare training are still promoting a continuum of care approach that is out of date and results in unnecessary moves and institutional care. This causes—actually it requires—people to be strictly categorized in order to get certain care in a particular setting or box. It is inimical to the notion that care is portable, and it too often produces almost self-fulfilling, too-long stays in skilled nursing or hospital environments where people are not dying the way they want.

We have much to learn from the Netherlands and the Scandinavian countries, where death much more often occurs in one's home or in residential assisted living settings. Death happens on people's own terms far more frequently than here. I foresee that Baby Boomers and their children will push changes towards our approach to end-of-life care. At least I hope so.

The hospice movement, not surprisingly, has grown on the same timeline and pace as the assisted living movement. Both arose from the same desire: more humane, more person-focused end-of-life care. They are both reactions against traditional long-term care and nursing homes. No other country has close to the number of nursing homes in total or *per capita* that the United States has, nor are they used elsewhere like here, which illustrates that something is not right in the U.S. in our end of life, long-term care system.

Most people want to die at home, surrounded by their loved ones. People do not want to be kept alive by so-called heroic measures that often extend pain, suffering, and life in unfamiliar surroundings. It is possible in the Netherlands for someone at the end of life to announce the day they will die, as if it is a social event: as in, the party will begin at noon, with a celebration of family and friends. Champagne will be served. And at 4 p.m., with medical assistance provided by a doctor, they will turn out the lights. Needless to say, it is a very different experience. Most of us know what a "bad death" is, while not enough of us know what a "good death" is. I believe insufficient attention is paid to this matter of dying with choice and dignity.

An issue getting more and more attention is affordability, and I believe the news is not as bleak as some report. Experts assert that one needs an *income* of at least \$70,000 a year to afford senior living, and that less than 2% of seniors have such income, which leaves millions shut out. This is a faulty premise. We need to consider assets as well as income when examining who can afford private pay long term care. We must also look at a *family's* income, not just that of an individual senior. Many seniors have children who can pitch in or even pick up the entire tab. Secondly, very few residents use income to pay for the two years of long-term care that is the average in our field. Even at a rate of \$75,000 a year for an assisted living community, a two-year stay is \$150,000. So, the real issue is: how many people in the

United States can afford a $150,000 long-term stay at the end of life? When we account for not only the income resources of the family together but also the available *assets*, the picture becomes more promising. At Sunrise, we did an analysis of the major metropolitan areas where we build and found that more than 50% of the senior population can afford our care when assets as well as income were considered. And Sunrise is not the cheapest provider; indeed, we are in the upper quartile of quality and cost. We also often had residents who were Medicaid-eligible, but the children did not want their mom living in a Medicaid bed in a nursing home, so they paid for her to live with us. There were also seniors who were "house rich" but living on a low fixed income. They often could sell or gift the house to a child or grandchild, who in return help pay for Sunrise's costs. There are actually many such solutions, demonstrating that affordability, while a challenging issue, is not as dire as many assert.

I do not mean to minimize the issue of cost in our field. Long-term care often entails at least three hours of daily care per resident, which makes "affordable long-term care" an oxymoron. Inexpensive long-term care simply does not and cannot exist. There is a reasonable temptation to look for ways to reduce the cost of care, but there are simply no magic tricks that will make these labor costs of hands-on care disappear. For example, you cannot build a facility so cheaply that the expense for care is no longer prohibitive for some; the building is a relatively small portion of the whole bill for one's stay in long-term care. Even if you *could* construct our buildings for 50 percent less, it would reduce the cost to the resident by about eight percent. The bottom line is, long-term care is about providing hands-on care and is going to be expensive.

Can technology change the calculations and democratize the costs and access to care? This is unlikely. If you ask, how will technology improve care for infants and toddlers up to three years old? the answer is virtually the same. Yet no one is asking that question.

Ultimately, this issue will have to be solved by public policy that recognizes that those in the lowest economic brackets are going to need some form of financial assistance to afford long-term care, and this will have to be provided by the government. The good news is that, with about half of the U.S. able to afford private long-term care, government assistance should be able to assist those who cannot. We in the U.S. are not the first people to confront this. Other countries have figured this out. Long-term care is hands-on and labor intensive. It requires food and utilities and quality housing and a stimulating environment. There is no way to provide quality long-term care cheaply in a way that is morally acceptable.

To ensure that mission, vision, and values become integrated into the fabric of a long-term care organization, we must, you must, make the time to talk about it. For example, at Sunrise we always opened and closed a meeting with a story about our mission. Whether the meeting is about budgets or policies or staffing, we consciously made sure mission is always included. Otherwise, mission is too easily overlooked, and your company may become driven by nothing but dollars and cents.

What does that look like in practice? Sunrise has well established principles of service, and everybody can recite them. We look for stories to share that highlight them. It might be an example of a team member, or reading a letter from a family—something that shows our principles in action. So, we tell stories that illustrate how choice, dignity, independence, individuality, and nurturing the spirit have been enabled. Once, we created a deck of cards that had each of the values printed on one side and practical examples of how to support each value on the other, and we played card games that required us to recite the principles and the examples.

Culture is not just handed down on a tablet. It is the culmination of a teams' habits. To build the habits and culture we want, we must share stories, and communicate those stories often, so they become a commonly known shorthand. And we cannot be deterred by the size of our organization. Sunrise has over 45,000 employees, but we still make sure they all are trained and steeped in the same culture and habits. Needless to say, a little technology can help—using technology and effective communication systems means everyone throughout the organization can receive the same information, hear the same stories, and cultivate the same habits.

The concept of mission has another application for those aspiring to leadership in this field. And that is, wherever you are in your career arc, focus on the special mission that is at the heart of serving seniors. It is impossible to overstate the importance of this. Educators in subjects like gerontology ought to emphasize the special mission, which is central to the senior living industry. For practitioners, this focus on mission will be a tremendous asset. The requirements of the field are very challenging. To have the calling to be other-oriented and motivated by mission will help get you through the many demanding moments anyone will face in this career.

Without imposing religious doctrine on anyone, I would note that the Bible defines the term "religion" only once, and it is defined as the care of orphans and widows in their distress. It is significant that most of the world's major faiths recognize that caring for seniors, particularly frail ones, is the greatest calling. Whether one is religious or not, I would encourage those in, or going into, our field to be comfortable speaking about their passion for this special calling. On the hard, tough days when you need to comfort a resident, work extra hours, or a situation goes wrong, or you must soothe an irate or despondent family member, this extra motivation, this second paycheck, may well pull you through it.

As a student, many careers interested me. During college in Washington, DC, I worked as a speechwriter, which can be a perfectly respectable profession. However, the end result simply did not satisfy me; I thought, all that work literally turned into nothing but hot air. There was very little left tangible at the end of the day. Yet when I volunteered in a nursing home with my mother, I had a completely different response. Just being a friendly visitor, to be surrogate "family" for a resident who really needed to talk, hold hands, and feel comforted, that was truly meaningful. And it drew me to this work. As did my experiences with my Dutch grandmothers. My decades in senior living, pioneering the Assisted Living movement, have been immensely gratifying.

I also happened to be so fortunate in meeting a man whom I have been blessed to work with and have as a friend for a long time: Victor Regnier. Victor is a world-class architect, educator, and author whose specialty is housing design for older people. We met by chance in Holland, not long after I started Sunrise. Our story illustrates the serendipity and importance of relationships: Victor not only helped Sunrise design many of our communities, but his being a professor and former dean at the University of Southern California led directly to the privilege I have had in writing this chapter for this book.

END

In 1981 Paul Klaassen started Sunrise Assisted Living to pioneer a fresh approach to long-term care. Their model introduced new physical environments and design as well as resident-centered principles of care and service which influenced how care and services were provided in those fresh, residential-feeling environments. The Sunrise Model, which they called Assisted Living, was based on the Dutch long-term care model which Paul observed in the Netherlands where his two grandmothers lived. Paul served as Founder and CEO of Sunrise for nearly 30 years. Sunrise grew from a single community in the Washington, DC suburbs in 1981 to nearly 500 communities with 50,000 employees in many major metro markets in the United States, as well as throughout Canada and the United Kingdom. Today, Paul continues to serve as an advisor to the Sunrise CEO and Board of Directors. Sunrise is still known today for its mission-driven culture and focus on resident Quality of Life, by emphasizing: Freedom of Choice, Protecting Privacy, Maximizing Independence, Celebrating Individuality, Involving Family and Friends, and Nurturing the Spirit. For more about the Sunrise legacy, read The Sunrise Story at https://www.sunriseseniorliving.com/about/company-info/sunrise-story.aspx

Chapter 7 Perspectives on Corporate Culture in Senior Care

Rick Matros

I graduated with a Master's degree from the first class at the University of Southern California's School of Gerontology, in 1975. It was the 1970's and I enjoyed being a hippie. It was an adventure to be in that first class with no particular plans for my career.

My first job was at a nursing home as its activity director, which was a lot of fun. Nursing homes in those days often played the role that assisted living does today. Most of the residents had their full cognitive abilities and few were seriously ill, so it was a very different nursing home environment than we have now. I believe I was the first male activity director in Southern California, and I loved the job.

Soon enough, the Facility Administrator asked me if I wanted to get my administrator's license, and especially since I had not studied business in graduate school, I accepted. The company was independently owned, with 10 or 12 facilities, and it was a terrific opportunity. When I received my license, they assigned me to a facility as the Administrator. Then three weeks after I received my license, the company filed for bankruptcy. Our cash flow stopped, and payroll checks bounced. I was only 23, had little experience in the real world, and had no idea what to expect.

Many of my fellow administrators in the company bailed; they did not want to be associated with a bankrupt company. When the health department visited the facility, officials basically ordered me not to leave my job; they effectively threatened me into staying, though I was not planning to leave anyway. I suddenly found myself with many new responsibilities, including using cash collections every day to go out and buy supplies for the facility and its residents. It was a strange situation, but I stayed there through all of it. Ironically, this experience made me realize that I liked the business side of long-term care.

R. Matros (✉)
Sabra Health, Irvine, CA, USA
e-mail: rmatros@sabrahealth.com

M. Lifschultz, E. L. Schneider (eds.), *Paths to Leadership in the Senior Living Industry*, International Perspectives on Aging 29,
https://doi.org/10.1007/978-3-030-53966-5_7

The owners asked me to work at their large facility in downtown Los Angeles, which was in a rough neighborhood. Indeed, the head administrator carried a gun. But I had always sought challenging situations and I perform well in adversity, so I took the challenge and did a good job. From there I joined a company that ran nursing homes, "board & care homes," and facilities for developmentally disabled kids and adults. I ran a couple of those facilities and loved the work. Ultimately, I became director of operations and learned the business from many angles. As I was gathering these experiences, I was definitely "paying my dues," though I did not think of it that way at the time.

From there, I joined the largest provider of long-term care at the time. Though I climbed the corporate ladder, it was the rare job that I did not like. I was exposed to the C-suite, which helped me grow. However, I was working for a company that required staff to dress and talk a certain way—to "fit the mold"—and that did not suit my personality. I wore earrings and my hair was long. Despite my persona, I was promoted because I happened to be very good at my job. However, talent and hard work can have its limits. I really did not fit in with the other executives and that can make work pretty dreadful. I quit. Not everyone in my situation would have made that decision. I had two little kids and little savings. But I felt like I couldn't breathe in that environment. I promised myself that if I was ever in a leadership position where I could impact culture, I was going to be the most anti-corporate corporate boss around.

I joined an extremely progressive company in the space. They were re-imagining the business. They were the first company that focused on high acuity Medicare patients and rehabilitation—getting folks out of nursing homes and back home with home care. The operators were forward-thinkers, and what they were doing later became the norm in the industry. For example, we had sub-acute units well before anybody else did and we had ventilator units in our facilities when other providers did not.

They hired me to be their president, which was a great opportunity, though it soon became evident that the owners were not managing the business prudently from a financial perspective. There was an atmosphere of risk and it was anticipated by some industry observers that the company was headed toward bankruptcy, because along with being bold and progressive in delivering services, they were also financially leveraged to the hilt. I joined the company knowing that there was some uncertainty, but I still viewed it as a unique opportunity.

Since I was hired as the president, I immediately visited all the facilities; I wanted to meet the administrators and staff, observe care firsthand, and get a sense of how things were run. Some of the staff, which included the woman I would ultimately marry, were quite skeptical of me as the new president simply because it *appeared* that I would run the company no differently than the current management that had put the company in jeopardy.

About six months after I joined, the company filed for bankruptcy. That, too, was quite a learning experience. We had to go to bankruptcy court every month and deal with the various parties, such as the creditors and debt holders. It was a tricky situation to navigate. While I was deeply involved in operations, the owners had strong egos and not only demanded to be the center of attention—relegating me in the back of the courtroom during the monthly bankruptcy hearings—they also resisted taking constructive steps, such as divesting and restructuring; the company was their "baby" and they weren't going to be told how to treat it. Meantime, when the creditors and the judges would ask for status updates and details, the owners would defer to me, and every time we went to bankruptcy court, the judge moved me up another row, until eventually I was in the front of the courtroom. Ultimately, the creditors seized control of the company and terminated management—except for me. They asked me to run the company.

We emerged from bankruptcy, got the company stabilized, and later merged with a newly public company, Regency Health Services. They were bigger than us, but we retained management and board control. We grew Regency nicely and within a few years, sold it to another public company in the space. At that point, after about twenty years, I thought it was time to leave the senior care industry.

I acquired a community-based company that focused on rehabilitation for traumatic brain injury victims. I delegated operations to a couple of my top executives, and they did a wonderful job expanding the existing facilities, building new ones, and growing the company. Simultaneously, I started a dental practice management company and ran that for a couple of years, raising capital and acquiring companies to form the platform business. After about two years, I promoted one of the executives to be CEO, though I did stay on the board of directors for another ten years, during which time we grew into the largest dental practice management company in the country, and actually sold it twice.

Then I decided to take time off to be with my family. At that point, about 20 to 25 percent of the players in skilled nursing went bankrupt because there were major changes to the reimbursement system that went further than the government intended, and they eventually increased reimbursement materially to assist the industry. By then, however, it was too late for many companies. We had seen the changes coming and believed it would be harsher than expected, which is why I had sold Regency. Companies were being acquired for very high valuations based on the expectation that the industry would continue to grow and the scope of impact of the reimbursement system change was underestimated. And as companies were filing for bankruptcy, I started getting calls from companies around the country, asking me to return to the business.

My response to these inquiries was always the same: yes, I will run your company, *if* you let me relocate your headquarters to Newport Beach! That usually ended the conversation, because for a company in bankruptcy, gaining stability was important. But the creditors of Sun Health, which was based in Albuquerque, New Mexico said, "Ok." That was really not what I expected or even wanted to hear. Though the truth is, I did miss the business.

After joining Sun, it took over two and a half years to restructure the company. It was an enormous challenge, because it wasn't just the finances that were a mess. The delivery of care was problematic. There were also rampant licensing problems. The culture was awful; we had an environment in which executives at the highest levels had tolerated bad behavior towards female employees for a long time.

We set up a system for employees to report wrongdoing without fearing that there would be consequence to their careers. I had to relieve people of their responsibilities it seemed almost daily. To staff up but also to change and improve the company's culture, I started recruiting people who previously worked for me.

We turned Sun into an outstanding company. We downsized by about 50 percent, divesting bad assets while maintaining its healthier core. We addressed the licensing issues and the care issues. Then we started rebuilding by acquiring some regional companies. Sun had always leased all of its assets, but after restructuring, we acquired companies that actually owned the real estate, and we ended up with about a 50/50 split of leased and owned facilities. When a recession hit, though, we lost our ability to grow through raising capital in the public markets. This caused me to explore other means of creating value for our shareholders.

Healthcare real estate investment trusts (REIT's) started growing in the 90's. I was studying the possibility of creating one and had the opportunity to join the board of a REIT run by a friend. This enabled me to deeply understand the business. I came to believe that it was a viable option and presented that concept to my management team and to my board. The idea got the green light, and about a year later we had created Sabra Health Care REIT.

The core issue that led to this decision was this: I realized that although our company's stock had reached a new peak, its market cap still did not reflect what I believed was the true value of the company. The market was valuing either the business or the real estate, but not both. Senior housing has always been a real estate-driven business, while skilled nursing has always traded on a multiple of cash flow because the business is so complex. Now there is more attention paid to the value of the real estate in skilled nursing, but that hadn't been the case at the time.

We decided to take action intended to capture the value of *both*.

We separated Sun into two separate publicly traded companies. One entity would be the new Sun operating platform with their leased assets and ancillary businesses. Sabra would own substantially all of Sun's currently owned real property portfolio and operate as a real estate investment trust under the name Sabra Health Care REIT. The day that we did the spinoff, the two separate companies were worth twice as much as Sun's value was the day before. And this was despite the increased risk inherent in Sabra's having Sun as its only tenant; if Sun were to have trouble, Sabra's value would suffer too. We needed to diversify away from the one tenant, by adding additional skilled nursing *and* senior housing. That's what Sabra has been doing since. Sun was acquired a couple of years later, and I remain the Chairman and CEO of Sabra.

Looking back, it was my boss at Catered Living, Jim Otto, who had the greatest impact on my career. He took me under his wing and made me director of operations at a relatively young age. He was smart and charismatic, and he really cared about the patients. He had a great instinct for the balance that we must strike in this industry—between providing care and making profit. His approach was, the best way to have a great business is to provide great care. He had a strong ego, yet he didn't impose it on other people. He helped me understand that I wouldn't get lost as long as I had faith in myself. He helped instill the sense that I could aspire to do things and do them my way—that there wasn't one way of doing things. You could do the right thing and be true to yourself, and not be intimidated by the forces around you trying to force you into conventional behavior.

I have always been hardworking and have achieved a high level of success, but my career has always been, and continues to be, fun. A big part of that has to do with culture. I like to talk about culture, because I think it is *the* most important component of any organization's success, and it was always my biggest challenge as the head of nationwide service companies, where at times we've had as many as 30,000 employees.

At the local level, you want leaders to imprint their values on the business because they're running the show there. However, it has to be consistent with a company-wide culture. It is a tricky challenge to define a corporate culture that is consistent in every location yet also has a local character. An advantage of running a leaner operation, like I do now with Sabra, is being able to have the culture exactly as you want it; being able to do everything you've always wanted to do for your employees, which you can do with 30 employees but you can't always do when you have 30,000, especially in terms of pay and benefits.

Along the way, we didn't achieve certain things that we wanted to. But I prefer to view such things philosophically; in every war you're going to lose some battles. Another way of saying that is, I tend to focus on the overall journey, not every bump in the road. I've always been fine with mistakes or things not turning out as well as expected, so long as we own up and learn from them. It's also important to recognize that certain things are not under your control and the most thoughtful processes may not always yield the optimal results.

In my first job as an administrator, when I young and inexperienced, I unwittingly joined a company that soon went into bankruptcy. It was especially tough because I was young and idealistic, and felt a deep commitment to the patients and their families; I did not want to abandon them. But I was very inexperienced, and really had no frame of reference for dealing with such a fraught situation. Making a commitment to stay on, though, was the right thing to do. I trusted my instincts. It set the course for my career.

As previously noted, I faced challenges every day at the company where the culture was terribly stifling; where management wanted to keep people in boxes, requiring us to dress a certain way, to conform to how they wanted you to be. The environment caused me to feel that we employees were not bringing value, that we were basically commodities. That really shaped me in terms of how I approached setting culture as a leader. The fact is most of those that work in our industry are good, compassionate people who believe in the mission of caring for our elderly.

The key is the culture that you develop, and that goes hand-in-hand with the uniqueness of our industry. In our business, of course you have a direct impact on the physical and mental well-being of people, so there's a passion and commitment that comes with that. I have always found that the people who work in our industry are deeply committed.

Of course, nothing has tested our collective industry, as well as the leadership, culture, and commitment of individual organizations, the way the COVID-19 pandemic has—and will continue to. Having been CEO of several organizations through several decades, I have led organizations through crises before, and I rely on both my experience and instincts to guide my plans, actions, and decisions. I believe that the following is an effective approach for this unprecedented crisis, though it can be applied to other crises, large and small.

Above all, leaders must maintain a sense of calm. In a crisis, an organization's "nervous system" tends to get its signals from the top. A calm leader will likely keep an organization calm. And care requires calm.

Indeed, care for the safety of your residents and staff is, as always, the highest priority. In a crisis where lives are on the line, the next priority is peace of mind for the families; leaders must be sure to demonstrate and communicate to families that care of their kin is being well-managed.

Communications must also be directed toward your staff. Ensure your people are continually hearing what is happening in the field. Emphasize the role of the staff in your facilities, because they are heroes and you want everyone to see that.

As you look at your overall operations, be thoughtful in your approach to any necessary changes. Be flexible and willing to relook at how everything is approached so that when you make changes, such as moving to WFH (work from home), you don't lose effectiveness. Be clear that we need to be prepared to utilize all our resources to assist where needed. In short, prepare your organization for change.

While the pandemic rightly became almost the sole focus of most organizations, including ours, in this and any crisis leaders should also find moments to point out that the strategic decisions made prior to the crisis positioned the organization to have the strength to make it through and come out on the other side as a healthy, strong enterprise. That understanding will ease the path for decision-making going forward and give staff the strength of optimism.

Finally, be cautious as you look ahead toward normalization. In the case of COVID-19, we learned much about the virus but also continue to learn how much we don't know. Therefore, resist the urge to rush back to business as usual. Be patient and let others jump back in as they inevitably will. Then learn from *their* experiences.

By taking this approach, I believe we were, and continue to be, able to project a sense of calm and thoughtfulness and thus provide our people with the confidence that their leadership can and will make the right decisions. When your people believe in their leadership, they will perform virtual miracles for the organization.

Other illustrations have come in recent natural disasters, such as massive floods and fires, from California to Texas to Florida. We had a number of facilities that got evacuated, and one that even burned down. Yet in each and every case, our employees stayed with their residents. So, as a leader, you must create an environment that makes employees feel good about their commitment, makes them feel like they have value, and makes them feel they are contributing to the mission. As a leader, you don't want your people to think that *you think* that they're just a commodity, there just for a paycheck. That starts with culture.

Wherever I have been a leader, I have set out to create a strong and distinctive culture. A big part of it is giving people a voice. I have always visited facilities so our people could talk to me, and I invite them to email me; it's not every company where employees can email the CEO. We also put structures in place that recognized those voices, such as having regional committees—comprised of the local administrators and the directors of nurses—that had the power to approve or reject any policies or procedures that corporate HQ wanted to institute. And we have used that model to encourage our executive directors to do the same thing within their facilities: to let people have a voice. That is the key.

Another deeply important aspect of culture is having a very clear mission statement that reflects the core of your cultural values. At Sun, for example, we conducted a process to get input from many employees and it resulted in a mission statement that "Caring is the key of life," which borrows from the title of a Stevie Wonder album. It accompanied our logo; it was on every piece of paper. By contrast, I once worked for a company that printed its mission statement on the back of business cards, and there were six or seven points. The CEO would visit sites and ask people to recite them from memory; if they did, he actually gave them a small cash prize. As CEO, I didn't want to have to pay employees off for memorizing our mission statement. We also had a strong sense that having a clear mission statement about caring would take care of itself, in that it didn't mention the actual business.

You can talk about culture all you want, and you can even put policies in place to encourage culture, but I have found that you also have to find ways to make it happen. While it may be counterintuitive, putting formal structures in place actually facilitates the culture that we want to create. At Sabra, we have created an environment that welcomes smart people and strongly encourages them to provide input, to speak their mind. Now, this is not for everyone. Even smart, talented people can prefer a more prescriptive environment, where they know precisely what their job is, what's in their silo; and they execute accordingly. There is nothing necessarily wrong with that. But for my work style, I've needed people that were more maverick in their mentality. Yes, they need to understand that certain policies and procedures must be adhered to, but at the same time they need to be comfortable creating their own initiatives. Different kinds of people thrive in different kinds of environments, and as a leader you need to identify the kind of people that you like to have around you.

I have always looked to create a culture that accommodates assertive voices and restless minds. I would have a sign on my desk that said, "If it ain't broke, break it," because I believe that as soon as you feel like you're doing something right, it is easy to slip into complacency. But you cannot assume that just because things have been working, they are *always* going to work, simply because the environment is always changing around you.

Indeed, change is coming to our industry in many ways and on many fronts. For one, there is good cause to be concerned about a shortage of workers for our industry. The occupation "Health aide" will soon be the largest in America. It is estimated that the need will be for five to six million such workers, yet there will potentially be a shortage of 1.5 million. In addition, historically there have been shortages in nursing and therapy. We also face a shortage of geriatric physicians, as this is one of the lowest paying specialties in medicine.

Some operators are able to recruit care workers from abroad, but that is also dependent on a conducive visa program and immigration policy. Historically, many of these workers have come from the Philippines, India, and a number of African countries. One of the reasons it has worked so well is that companies would attract whole families, in part by paying for their housing and effectively importing a community, which engenders deep loyalty. But to sustain immigrants as a key source of workers will require the support of policies from the federal government.

Another aspect of shortage will manifest in skilled nursing, where there already is a declining supply and an increasing demand. It has been estimated that by 2025, every skilled nursing facility will be 100 percent occupied. These facilities are taking more post-surgery patients, while "traditional" indigents on Medicaid who need long-term care are not being taken by skilled nursing facilities anymore. And they are not being taken by senior housing because it is largely private pay. This leaves the mom-and-pop skilled nursing providers to accommodate them. However, the viability of these facilities will be a struggle, because they typically do not have the wherewithal to change their model from being an old-fashioned nursing home to being a step-down unit from the hospital. Supply will, therefore, be choked off as these operators go out of business.

In the last 15 years, we have seen a decrease from 23,000 nursing homes to 15,000 nationwide. So along with a lack of caregivers for frail older persons, there will be a lack of *places* for them to go. Without a dramatic change in the trajectory of this particular industry, there will be, literally, blood on the streets as many in the truly indigent elderly population will have nowhere to go. Unfortunately, the government will not react until that suffering is right before our eyes; we should not expect a problem that is ten years away to be addressed currently in our political system, which has a two-year focus.

Senior living, which is not an old industry to begin with, has already changed a lot, evolving from a hospitality model to a needs-based model. Assisted living and memory care today are basically "nursing home light," acting much like nursing homes used to. However, that is happening primarily on the private pay side. That model will continue to evolve, and assisted living will become more and more of a medical model. Independent living has changed less, though operators now have arrangements with home health agencies to provide supportive services. The more senior housing becomes needs-based like skilled nursing, the more economically robust the space becomes.

Developing models for less affluent seniors may depend on Medicaid waiver programs. Some states have such programs, and this may enable senior housing operators to specialize in the lower- middle and middle-income populations. Of course, each state has its own Medicaid rules, and some waiver rates are so low that they are not viable as a fix for the problem. We don't invest in senior housing properties that accept Medicaid waiver patients because we don't trust in the longevity of those programs. There are operators that do this, which is good to see, but their financing sources are limited. The need is indeed great, but if the programs don't get stabilized—which is a condition for more dependable financing sources—then it is hard to see the quite small percentage of operators who take Medicaid waiver patients today expanding dramatically. Commercial insurers will play a role in senior housing and can be part of the solution.

The primary change on the business side of senior living is that is has become far more complex and sophisticated. To take a critical example, government funding used to be totally dependable and occupancy was always high. Thus, by controlling costs and providing good care, one was effectively guaranteed a prosperous business. By contrast, today the dynamics are unique to every market: clinical products and reimbursement models are more complex, and competition is wider and fiercer, in no small part because it includes the powerful health systems business. So, as a baseline, to be successful in this the industry, individuals have to be far more educated than was previously the case.

The complexity of the industry also demands that companies excel at decision-making, and for that there is nothing more important than being *disciplined.* Being a REIT enables us to be an opportunistic investor, which is critical to our success. Keeping extrinsic, or outside, forces away from decisions helps a lot. Decisions should not be driven by prescriptive considerations; that is, by how things *ought* to be. For example, long-term plans, which never survive contact with real business cycles, can confine your thinking and box you into unwise actions. They are the enemy of opportunistic decisions.

Similarly, when you are publicly traded, Wall Street watches your decisions quite closely. It is crucial not to create expectations—which are also extrinsic—that you may not deliver on. Recently we had a year in which we tripled our size, but the year before that we did less investing than we'd ever done. The actions were diametrically different, but the discipline and its message to the market were consistent: We are not going to grow just to grow; we invest when we see opportunity. That takes discipline.

Being disciplined, however, does not mean that you can't be creative. To some, those words may sound like opposites. They are not. They can, and must, co-exist. Enabling your people's creative thinking is straightforward: encourage it, provide mechanisms for it, and foster an environment where speaking one's mind is safe and comfortable. Creative thinking generally implies constructive criticism of existing ways. Even these conditions may not always yield everyone's creativity. Factors such as how someone was raised with respect to authority and past experience with corporate cultures may well interfere with the process, especially if one is addressing the CEO. I always try to pay attention to getting the person or people in front of me to know that there will be no consequence for this type of input. Our headquarters staff is quite small, and we gather most Friday afternoons to have a wide-ranging session to commiserate or celebrate the past week's events, inside and outside the company.

Cultivating creative input posed a bigger challenge at my previous company, Sun, because there were thousands of employees. To keep in touch with them, I very actively involved HR. In many companies HR is largely administrative, tasked with matters like managing employee benefits and compliance with labor laws. But HR can also be a driver of culture, including: putting those mechanisms in place for people to have sounding boards and give input; doing succession planning; creating career ladders, and more.

In most companies, the senior HR executive reports to the chief administrative officer, the CFO, or the COO. I always had HR report to me directly as the CEO. We would wall off the administrative matters, but when it came to culture, I wanted our people to know that I was directly involved, which sent the message to the organization that this stuff is *important*. Moreover, I believe that facilitates a culture that honors the value that people bring to the job every day.

There is no room in our culture, or our industry, for hubris. Hubris is the main cause of mistakes and bad judgment. There is a popular street idiom that is too coarse to repeat, but let's just say that when people start to believe they personally always smell like roses, it's a bad sign. I tell this anecdote to contrast hubris with humility: I was at a conference and was asked by another CEO why I was having a staff dinner rather than joining the CEO's dinner. I said, "Our staff is the reason I'm CEO!"

It is somewhat surprising to find hubris in this industry because there is so much about it to humble you. When you visit facilities and see residents and patients and their families, how can you help but to be humbled? I recall visiting an Alzheimer's unit at one of our skilled nursing facilities and speaking with a woman who was visiting her husband there. He had no idea who she was, *and* he had a girlfriend in the building. Yet this woman was bringing food to her husband and his girlfriend every day.

Considering that we routinely see situations like that—stories of joy and suffering, exuberance and sacrifice—I look around at some of my colleagues in the industry and think, do you see what we're *about*? Yes, success has its rewards, and, yes, it requires an ego to rise to the top, but this does not mean you are better than anyone else. But if you think you are, you will make mistakes, period.

It comes down to conducting an ongoing self-assessment: no matter how well things are going, you must ask yourself, what could you be doing better? There is always something that you could be doing better. In addition, you have to be getting input from others, constantly and at a level that it is meaningful.

My wife reminds me of this all the time. One of her favorite jokes is, it is a good thing you know how to run companies, because if you didn't, you'd need 24-hour care. Which is probably true, and which is why I'll never retire—because I don't want her to say, see, I told you so.

END

Rick Matros has served as Sabra Health Care REIT's Chairman of the Board, President and Chief Executive Officer since its inception in 2010. He was Chairman and CEO of Sun Healthcare Group, Inc. from 2001 until the transaction that created Sabra in 2010. Mr. Matros founded and served as CEO and President of Bright Now! Dental from 1998 to 2000, remaining a Director through its subsequent sale in 2010. In addition, Mr. Matros was a Principal and Partner of CareMeridian, LLC, a company specializing in traumatic brain injuries, from 1998 until the sale of its operations in 2006. From 1988 to 1997, Mr. Matros held CEO and other executive positions at Care Enterprises and Regency Health Services, both publicly held post-acute companies. Mr. Matros is a member of the Board of IsraAID US, a lay leader in the Jewish Community, an active member of the National Board of Directors for AIPAC and he sits on the Governing Board of Aleh-JNF. He is also the Executive Producer of Sabra Films, LLC. He is a die-hard fan of the Los Angeles Lakers and the USC Trojans.

Chapter 8
Creating a New Paradigm for Senior Living

Patricia Will

It all began in the early 1990s when my mother-in-law developed dementia. Where would be the best place for her? Nursing homes had corridors filled with rows of dementia patients strapped in wheelchairs waiting to die. Physicians in psychiatric hospitals regarded dementia as a normal part of aging and were not interested in treating these patients. Home health care did not have staff trained in the special needs of patients with dementia. So, though I had a successful career as a real estate developer of medical properties, I became interested in creating another option for people like my mother-in-law. It was clear to me that a void existed in the care spectrum, and I set out to create the right product to fill it. My dream was to create a *place*, not only where I would want to see my mother-in-law and others reside, but where eventually I would be glad to live. I spent several years researching the state of the industry, conceptualizing both what to create and how to bring it to life. This preoccupation ultimately became my second career. The result was Belmont Village Senior Living, which I co-founded. Today, we have over 4,000 units in approximately 31 communities from coast to coast, and in Mexico.

Many people had an impact on my development as a leader and helped shape the contours of my career. Two mentors stand out above the rest. They impacted my development in distinct but complementary ways. First, my mother, who was a careerist and an entrepreneur when few women were either. She raised me to believe that I could be anyone and do anything. She inspired me to realize that being female was no limitation, that I should just *go for it*. She instilled in me all the confidence I would ever need. This has always served me well. My self-confidence was especially valuable when I went to work for my other great mentor, Walter Mischer, Sr., a powerful entrepreneur who, with no college education, became a major Texas land developer and business giant. When I interviewed for a job with him, there were few

P. Will (✉)
Belmont Village Senior Living, Sherman Oaks, CA, USA
e-mail: pwill@belmontvillage.com

M. Lifschultz, E. L. Schneider (eds.), *Paths to Leadership in the Senior Living Industry*, International Perspectives on Aging 29,
https://doi.org/10.1007/978-3-030-53966-5_8

women at his company or in the real estate development field. His first words to me, which illustrated his *modus operandi*, were, “So, what the hell did you learn at Harvard Business School?” It would have been easy to be intimidated, but I believed in myself. While Mischer appreciated my formal training, he taught me how to think, react, and negotiate without relying on a pile of fancy spreadsheets. Self-confidence, rigor, and curiosity; these are the qualities that have shaped my work life.

Though being a developer of medical real estate and a builder/operator of senior housing may overlap, there are key differences. Real estate development is driven by projects and transactions, while senior living is focused on *people*. Our industry has immense capital requirements and operating risks, but what concerns me the most is not occupancy rates, profit margins, or liability—it’s recruiting and retaining terrific people who can work effectively in teams to serve our residents. Even the finest business schools cannot impart how crucial the dynamics of human behavior are to organizations in people-centered industries such as ours; this must be experienced to be learned, and I believe this is where leadership counts most. Ultimately, our industry requires the seamless efforts of so many different types of people—nurses, doctors, chefs, nutritionists, technologists, managers, financial analysts, dishwashers, van drivers, *and more*. We must be able to find and retain and excite and motivate all of those people and get them working as a team. People are your chief competitive advantage. Inanimate buildings—no matter how humanely or beautifully designed—need to be brought to life by your people in order to best serve your customers.

We devote considerable time and energy to the mysterious alchemy of teamwork. That process starts by encouraging our staff to be in contact with our residents and with one another. Belmont people, even or *especially* our leadership, including me, should not spend much time sitting in offices. We are constantly in our communities. For example, each year I personally have dinner with approximately 1000 Belmont line employees.

Having a close-knit, supportive culture is crucial in senior living. This business is extremely demanding on a personal level. Taking care of people’s lives is rewarding, but it also can take its toll. Obviously, providing care is hard work. Even harder, though, is the inevitable loss and grief employees routinely experience since they are working with, and often become quite attached to, an older population of residents, many of whom are frail and will die. One way we recognize this inherent stress is by sharing the financial upside with leadership and others who contribute to our success. As a result, our people are incentivized to help create lasting value through their work. We also celebrate successes on a non-stop basis, rewarding outstanding employees with recognition for their work. We know one another other very well, and we cultivate a sense of family.

Belmont Village’s familial culture starts with our hiring and training practices, which have changed fundamentally over the years; we no longer rely on outside search consultants and employment agencies. Instead, we have a dedicated group of internal recruiters and trainers, all of whom have previously had operating roles with Belmont. This team does nothing but recruit and train, and they do it effectively because they understand what it takes to be successful within our organization. We have been honing our in-house training program for years, and we put management associates right out of school into the program. We deliberately select mentors and deploy company leaders to nurture these new employees. Our "graduates" are among our most successful leaders.

Your employees should derive satisfaction out of their work. You usually find out very quickly who does and does not. Working with elders and their families and becoming extensions of the families themselves is gratifying. This is true whether you are a Harvard MBA working in finance or a line driver shuttling residents to activities. This gratification is why our employees stay with us. I certainly feel that way. This is one of those rare businesses where you can do well by doing good.

Moreover, the prospects for career-minded, aspiring leaders is exceptionally bright in our industry. There may be no business whose prospects are brighter than ours when you consider the demographics of the next 30 years. The industry is now at the cusp of experiencing massive and sustained growth. As 70 million Baby Boomers reach old age, our industry will be transformed. It is mind-boggling to contemplate the demand that will be placed on seniors housing and long-term care, given the fact that every day another 10,000 Baby Boomers turn sixty-five. And our industry will be growing regardless of the economic cycle, as the need for housing and services for older adults will naturally persist.

Many industry leaders believe that the arrival of Baby Boomers will be a game changer, in that it will render many aspects of our industry obsolete. I don't subscribe to this view. Boomers lived in college dormitories, fraternity and sorority houses, some even on communes; then they moved to cities and had roommates. They accepted congregate living as young adults. This generation will therefore be very familiar with and more likely to be receptive to moving out of single-family homes and into senior living communities. Baby Boomers will actually be more willing and participative customers than those we have now. In contrast to the Silent Generation we are currently accommodating and who are known for their conformity, the Boomers will be more likely be active partners, letting us know what they need in terms of services and environments. Many in the industry disagree with me and predict an industry that will morph into one that would be unrecognizable to us today. Others predict a trend toward "aging-in-place," with more Boomers opting to remain in their homes. Boomers love to socialize, so I remain bullish on the appeal of our product.

One change I observe is in our workforce. Some industry leaders are concerned about the work habits of the Millennials, who are becoming the majority of our caregivers. My experience of Millennials is at variance with this view. We cultivate, honor, and love the large number of Millennials in our organization. We appreciate that they are purpose driven. While they want careers that will pay the rent, they also deeply care about personal satisfaction and mission. This is terrific for us Baby Boomers, not only as employers, but also as future residents for whom the Millennials will be providing much of the care. Unfortunately, though, there are not enough of them. We need to go all out to attract them to this terrific industry.

Another looming challenge for our industry is the issue of affordability. Private pay senior housing is *expensive*, and the subsidized nonprofit side is limited by financial qualifications and declining government support. In California, for example, where we have operated successfully for almost 20 years, the realities of the market are that you cannot acquire the land, create a community, and employ the many professional hands needed to responsibly deliver a quality product *and* still be able to make it affordable for less affluent seniors, whether in Los Angeles, the Bay Area, San Diego, or coastal Orange County. Of course, that is not the case for every state, city and town in America, though it certainly is in and around many major metropolitan areas. Part of the solution is delivering our products and services elsewhere. This is why Belmont has already entered a foreign market, namely Mexico.

I anticipate developing many communities in coastal Mexico over the next ten to twenty years, for Americans. My strong sense is that many of us Boomers would be very happy having seniors housing and care with a margarita and a view of the sea.

I have spent almost half my life in Mexico. Though I was raised in New York, my stepfather is Mexican, and growing up I spent every summer in Mexico. I speak the language and I am very comfortable there. That is, in part, why we built a community in Mexico City. The community is highly distinctive, being part of a mixed-use complex attached by a skybridge to Mexico's only JC05 hospital (Joint Commission, level five tertiary care). This community takes care of frail, elderly seniors who are Mexican and Latin American nationals with adult children living in or near Mexico City. And we have plans underway in Mexico for future communities. Ultimately, having developed a potent operating organization in Mexico, we believe there will be a major opportunity to develop coastal resort communities there. Mexico will afford us the ability to deliver a more affordable product in a very pleasant setting with access to terrific first-rate healthcare, all just a few hours away from the cities in which we already operate, such as Los Angeles and Houston. It will not work for everybody, but it is going to be part of the access/affordability solution.

We are developing our operating skills, our programs, and our leaders of the future there. Care and insurance in Mexico is excellent. Americans who live in Mexico are able to buy insurance very affordably, because healthcare is so much cheaper. Common surgeries, such as hip and knee replacements, for example, are available for less than half of the US cost from first-rate physicians in great hospitals. Another advantage to developing senior housing in Mexico is its large, nurturing, and hospitable workforce. At Belmont in Mexico, we have many nurses available to us, and we have deployed MD's there in the same capacity as nurses in the U.S., thus adding quality medical care, while still delivering services that are far more affordable than in the United States.

I may be the only one in the entire industry with this point of view. Some of my industry peers call our Mexican effort "Patricia's folly." But we are working hard to prove them wrong. We are learning a lot and having fun.

If the most important decisions I make are about people, then a close second are the decisions related to what we *won't* do. Being able to say "no," rather than chasing every opportunity, is a key to success in our industry. As a leader, you must put a "Proceed with caution" label on any big new idea, because once you put an organization in gear to accomplish a goal, you are dedicating considerable resources and energy that by definition cannot go somewhere else. A leader must be able to recognize an opportunity but also be able to turn one down.

However, that does not mean that you should restrict creativity and bold thinking. In fact, I believe it is part of a leader's core responsibilities to promote creativity, to support and nurture it. Our ongoing Bright Ideas Program invites all Belmont employees to propose actionable improvements with an informal, back-of-the-envelope business plan. Typically, we get a couple of hundred submissions each year. The winners must be ideas that can be implemented within the year after being selected. A group of finalists are chosen, our corporate office helps them make professional presentations to the judges, and we implement the best two. The process is energizing and promotes company-wide thinking and learning while putting torque behind idea-generation and problem-solving. It is also a great source of future leaders. It could be considered a mash-up of some of the best practices found in the venture capital industry: a business plan competition and an internal incubator. Often the important ideas that emerge generate a response of "Why didn't I think of this?!" I believe you can coax or prompt creativity out of people, that it can come from anywhere, and that the best ideas come bubbling up, not from the top down.

A great example of a winning idea, which could probably be helpful to *any* company, was based on the near-universal human experience of finding that job descriptions so often fail to provide the content and the context for the actual work involved; you get a job and realize, I didn't think it would be like *this*! The elegant solution proposed was, make day-in-the-life-of videos for every position, showing in detail what a person in every hired role typically does on the job. What a fabulous and informative approach to showing to a potential hire, in a dignified way, that as a caregiver you may have to change a diaper. Or just how much fun it is to work as an activity coordinator.

Another area that we emphasize is leadership development. Nobody is born a leader. There is no magic recipe that yields a great leader. We come to leadership in many ways. I half-joke that I was out of new things to do and interesting hobbies, so I started a company. With no employees, and then only one assistant, and later a core group of people, I became the presumptive leader. That experience may be quite different from someone who joins a company and works their way up through the ranks and is recognized for their intelligence, values, creativity, and ability to inspire others. From my perspective, the issue is not how I became a leader, but how I was able to remain one.

It is not uncommon for entrepreneurs to start companies then fail to make the transition to managing the large and complex organization they created. The very gifts and skills that drive the entrepreneur to will a business into existence do not always translate to leading people. You need to share responsibilities, accept other's failures, spread the wealth that is created, and engage as a team member, even if you alone ultimately have to make the tough decisions.

Often organizations outlive their founders and are unprepared for that moment because post-founder leadership was never addressed. Given the stakes in senior living—we have people's lives in our hands, after all—this is not an option. Having discussed this with fellow founders in the senior living industry, we agree that you must create the environment and orchestrate the conditions wherein your successors can thrive as leaders. Fortunately, I found Mercedes Kerr and was able to convince her to leave a large public REIT to become President of Belmont Village.

I have always encouraged those considering careers to look at senior living. It brings me joy every day to see the lives that have been transformed by Belmont Village. We see seniors who were living alone in their homes, isolated and depressed, now excited for the day to begin so that they can interact with their new friends and lead or take part in activities such as book clubs and water colors. Finding a place for my mother-in-law has led to a wonderfully fulfilling career.

END

Mrs. Patricia Will co-founded Belmont Village in 1997 following a three-year research effort. The goal was to create a senior-focused community that reached beyond the existing paradigm in Assisted Living to create a comprehensive and unique space and service product. Today Belmont Village is a fully integrated owner, developer and operator of first-rate senior housing and care communities. The company is known for built-to-purpose developments, best in class hospitality and proprietary resident programming. University collaborations and partnerships with healthcare institutions are a hallmark of program development at Belmont. The company's $2.5 Billion real estate portfolio has capacity of more than 4000 units and employs over 3800 people across eight states and Mexico. The company's buildings and spaces have received numerous architecture, construction, and design awards. The company's associates voted the company as a Great Place to Work® in 2018, 2019, and 2020, since *Fortune* initiated its ranking for Aging Services. Prior to Belmont Village Mrs. Will co-founded Mischer Healthcare Services. The company served as the launch pad for Belmont Village Senior Living. Mrs. Will has served on the Board of numerous for-profit and non-profit entities. She currently serves on the Board of the USC Davis School of Gerontology and the UCLA Center for Longevity. In November 2018, she was inducted into the Texas Business Hall of Fame. Mrs. Will earned an M.B.A. with distinction from Harvard Business School and a Bachelor of Arts degree, Phi Beta Kappa, from Reed College, Portland, Oregon.

Chapter 9
Leadership in Senior Living: Serving People

John Woodward

I attended graduate school for business and hospital administration at the University of Chicago. After the two-year program, I joined a consultancy that advised companies in healthcare. It was a rather arduous "baptism." Our specialty was servicing hospitals that were planning major capital programs. Since the hospital industry is highly regulated, there were many technical matters to master, and *that* you don't learn in business school. The work also entailed intensive client contact, which included interactions with both management and boards. I gained considerable exposure to how organizations develop and pursue their goals, and I observed many approaches to leadership.

Over the course of three years, I worked with dozens of organizations. It was a rigorous education because each hospital had its own corporate culture, geography, board dynamic, and management style. I also learned about the various products and services of the healthcare industry. Then some colleagues and I started a healthcare finance operation for a small investment bank, which over the next five years grew substantially through about 30 mergers. Indeed, we grew so rapidly that we never had current business cards.

Healthcare investment banking was relatively new to the West Coast, where I was based, so there were vast opportunities to develop professional relationships and introduce new financial tools. Today's senior living industry may not be quite as wide open, but there are still parallels, in that the field is ripe for innovation and rapid growth. Although I thought at the time that I would work in banking for just a few years, I ended up doing it for about 35 years, working with hundreds of nonprofits, including hospitals, HMOs, senior living organizations, and skilled nursing facilities, across some 40 states.

J. Woodward (✉)
Front Porch, Glendale, CA, USA
e-mail: jwoodward@frontporch.net

M. Lifschultz, E. L. Schneider (eds.), *Paths to Leadership in the Senior Living Industry*, International Perspectives on Aging 29,
https://doi.org/10.1007/978-3-030-53966-5_9

As an investment banker, I learned about leadership from a unique perspective. The investment banker's work entails devising ideas that leverage a client's business, then persuading management and the board to follow your advice. The stakes for your clients are high, whether you are helping them chart a course for growth through a merger or expand via a major capital project. You are listened to, and to be most effective you must understand not only what they are doing but *why*. Your ideas must serve their mission, not just their margin. Therefore, as an investment banking executive, you expend considerable effort in trying to help lead your *clients'* organizations. At the same time, you must lead the people you work with in your own organization in order to succeed.

From this, I drew two meaningful lessons on leadership.

First, the ability to define and reinforce culture is what drives strong leadership, and as important is the ability to be sensitive to people's needs. Working nationwide, I learned that while strong similarities exist among people wherever you go, cultural differences and personal views vary. Thus, one must pay close attention to deeply understand the particular needs of those you lead. Fundamentally, you have to understand their challenges and offer them solutions. When you do this, almost everyone will enthusiastically accept your lead; when you don't, you will find your leadership undermined by indifference if not outright hostility.

Second, leadership means serving people, whether they work for you or you work for them. Period.

I was fortunate to learn these lessons early in my career. When your end product is a faceless financial instrument, it may be tempting to almost entirely ignore the people side of your business. Indeed, I observed many bankers dismissing an obvious conflict by focusing on creating those financial instruments that maximized profits for their *own* firms, rather than putting clients' interests first. This was awfully short-sighted (and also unethical). I realized, however, that devising the end product was just a small part of my job and that personal financial success would inevitably come if my priority was meeting my clients' needs. My mission-critical role was to help people define and attain objectives; the financing was a means, not an end in itself. The value of this lesson to leadership cannot be overstated.

I took this insight from my banking career to my current work as an operating CEO. One must be ever mindful of the broad perspective—what is colloquially called "the big picture." But you cannot gain much by focusing on the view from 60,000 feet if it means losing sight of what is happening on the ground. For example, in senior living if you think you are principally in the real estate business, you will fail your residents, their families, the community, and, most likely, your funders and investors. There *are* buildings to manage, but they will have no economic or social value without people *gladly* paying you to live in them. At Front Porch, we say and mean that we are in the business of meeting needs, not providing products and services. This is not merely a nuanced choice of words; it translates into a significant guide for how we act. Our Mission is "Meeting needs through excellence in 'human serving.'" Our Values state, "Customer Satisfaction is our hallmark and is the highest standard by which Front Porch measures success."

These two priorities of leadership—setting culture and serving people—are reflected in Front Porch's other core values, Teamwork and Trust, which state, "We are one team working to better serve our residents, clients and the community. Only through working together in an environment of trust will we achieve excellence." Teamwork is a value and practice that is infused through every effort of our organization, from care delivery to problem solving to strategic planning.

Indeed, the COVID-19 crisis has reminded us of the value of teamwork and the importance of delegating responsibility. This has been particularly true for Front Porch, as we serve over 5000 residents in 38 locations. Our local community managers have led our response to COVID-related challenges, with support from both divisional and corporate office resources. Our strategy has been based on a tiered response, with our regional managers supporting local leadership and corporate staff supporting both layers.

Communications has been an essential component of our response, with weekly conference calls with all community executive directors to share strategies and identify resource needs, including personal protective equipment (PPE) and testing kits. We have also been able to shift staff and PPE between communities as needed to respond to quarantining staff receiving positive test results.

Crisis is also an opportunity for leaders to pursue change. In addition to learning how to address our own critical needs, we continued to communicate with other organizations to share both knowledge and resources. Discussion of potential partnerships accelerated during this crisis period as we increasingly appreciated both the short-and long-term benefits of collaboration.

Byproducts of a crisis include tension, anxiety, exhaustion, fear and loneliness, but a crisis also can move us forward. We will progress and hopefully be better prepared to address the challenges ahead.

And we will take on those challenges as a team. Notably, Front Porch does not have a dedicated job position of strategic planner. This role is too important for one person to hold, even, or especially, within the C-suite. Instead, we treat strategic planning as an organization-wide responsibility. We encourage discussions, debates, and even struggles, then we iterate and revisit and decide and sometimes repeat that process. If people believe they do not have their hand on the tiller of the organization, that they have *some* influence, they are not going to take responsibility beyond the narrow scope of their job. Another way to state this is that I worry about everything and I want our people to worry about everything, too. We make worry a team project.

The wonderful executive team at Cedars Sinai Medical Center in Los Angeles has been an inspiration to me for 20-plus years, and I have emulated their model teamwork in my role as the leader of Front Porch. They personify what a team ought to be. They completely trust and rely on one another and sit in close proximity within a modest executive suite. They develop and execute the organization's vision together, and they share an enormous pride in their work. They understand that they are not perfect, and they rush to take responsibility collectively and seek a solution when an issue or decision goes wrong. I believe deeply that it is this culture that has propelled and maintained their success; Cedars is perennially recognized as one of the finest hospitals in the U.S.

Teamwork is a force multiplier, at every level of our organization. As a Bain & Company article put it so well, "Teaming people with different skills and perspectives together almost always produces more output than the sum of those individuals acting alone. After all, two heads are almost always better than one." (https://www.bain.com/insights/how-to-manage-a-team-of-all-stars-hbr/)

Soon after I joined Front Porch, all managers and executives went through an innovation program that was developed at Harvard. On a weekly basis, we broke into small groups and engaged on various hypothetical problems, though not those from within the organization. It was a general problem-solving exercise that enabled us to think hard about how we create, and how change occurs. The lessons still reverberate within the organization. We learned about how to take on risk without undue fear of failure, and we did so *collectively*, with teamwork and trust. To this day, we approach innovation boldly, and we do it together. We also use a model not unlike Cedars, encouraging conversation when things are not done correctly, whether internally among employees or with residents and their families. Especially on difficult matters, teamwork and trust enables taking responsibility and being proactive, which ameliorates even the most difficult situations.

Indeed, as with so many aspects of leadership, communication is essential to fostering teamwork. My approach is focused on enabling team members to talk to each other, whether the conversations are based on disagreements, complaints, failures, successes, or *whatever*. That is how I see my ultimate role: as the facilitator in chief, finding ways to get everyone to work together. I know that if I'm effective, my work product may be the least visible of anyone's in the whole organization! The people who work directly with our client residents are the ones whose work really counts. If I have helped those people be prepared to meet our clients' needs, then I have helped put together a great team.

The nature of leadership has changed and will continue to evolve. This change is best understood in terms of how younger generations of people can lead most effectively. Modern leadership is less top-down than it used to be. It is more decentralized, spreading responsibility around the organization. It is no longer the responsibility of leaders to force their employees to conform to a mold or a model. Until recently, leadership almost exclusively was the province of males, and organizations often had one "patriarchal" figure who created the empire and called all the shots. Today, leadership is about leveraging people's skills and guiding them toward reaching their highest potential. As a leader you do that at the level of both the individual and the organization. You must be constantly adapting in order to maximize how people are deployed. This requires you to have a generous attitude, to view your leadership as a personal commitment to your people. It means fostering a work environment that is challenging, rewarding, and, fun. These were seldom priorities of leaders in the past. I also believe—as do many others—that leaders must create the conditions in which failure is not necessarily a problem, but is more an experience to learn from, as well as one that signifies that people are experimenting, innovating, and actively promoting change. It cannot be overemphasized: true leadership means managing your organization as a team. If you don't have a strong team then you as a leader have failed. Leadership rewards humility, not arrogance. When Front

Porch hired me, among the first things the board directed me to do was create my succession plan. One of my ongoing priorities is to make my role less critical through delegation of decision-making while emphasizing teamwork, and I believe I'm getting close to accomplishing this. That is a very different approach to leadership than we have previously known.

Still, leaders often perceive themselves as needing to take on a more predominant role than they actually do. They assume that they should be the only captain steering the boat, when they ought to give up the tiller as much as possible. This stems from that paternalistic model of leadership that hopefully we are rapidly evolving away from, in society and in business. Leaders too often fail to listen to others and to give others meaningful responsibility in the management of an organization. As a leader, I fail in this regard fairly often and as a corrective I actively ask myself, how can I do things differently to share more responsibility. Putting more trust in your people and taking the approach of, "call me if you need me" is good advice for any leader. I always try to keep in mind that the strongest management team relies on collaboration rather than just direction dispensed by the leader.

I am learning more about our operational needs, and I am seeking closer relationships with our residents. By doing this, I can better understand our business dynamics and how they change. In addition, I am always looking to spread out responsibility for strategic development within the team, and then stand back as quickly as I can. It is not easy to recognize or accept, but if I keep defining the issues and the solutions, everyone around me will respond, "whatever you say, chief, you're the boss!" However, remember that no one wants to tell the boss, "hey that's a stupid idea." Therefore, you need to encourage frank feedback, no matter how damaging to your ego.

Senior living as an industry generally regards itself as having many unique qualities that necessitate different approaches to leadership and management. The healthcare industry had a similar view of itself until the 1980s. However, I believe that our industry is not all that different from others, especially those whose purpose is serving people. But if one were intent on finding distinguishing characteristics, I would acknowledge that the demographic of our client residents—the average age in senior living is mid-80s—does tend to magnify the importance of certain factors within the provinces of leadership and management.

Executive-level stability is critical because disruption not only interferes with your operations, it also can have ripple effects as mid-level managers may leave if their direct boss does. For example, losing your CFO can wreak havoc on your entire finance team, and before you know it, you're dealing with significant financial complications for your Medicaid recipients. There is a huge premium on uninterrupted service up and down the organization. Of course, even if you have created a terrific working environment, new opportunities will come to highly skilled managers and they will leave. When these employees decide to move on, I focus on being supportive and celebrating their work. By cultivating that kind of environment—and the reputation that comes along with it—we will increase our chances of retaining and attracting great people, including those from other senior living organizations.

Retention of lower level workers is equally important. Turnover is very expensive in terms of recruiting and training replacements. But most importantly, your customers, the residents, *hate* it. Change is very disruptive to them. Residents get attached to the staff who are so closely involved in their lives on a daily basis. There are a few fundamentals that are key to retention. First, you need to pay as well as possible. Second, you must have an efficient hiring processes. In our industry, hiring can require considerable time. We are competing for employees who can go down the street to a hotel, restaurant, or retail store, get hired immediately, and get a paycheck by the end of the week. Third, we need to train them while demonstrating our commitment to their future.

Fortunately, we have a "secret weapon" for retention, and that is the residents, because the attachment works both ways. Residents develop warm friendships with employees, and employees develop similar friendships with the residents. That brings a stickiness that we cannot create with money and opportunity. However, one cannot rely on that stickiness for the long term, since perhaps the gravest threat to retention is the prohibitive cost of housing for workers, especially in major metropolitan areas. For this problem, neither the industry nor our organization has a promising solution.

That issue helps illuminate perhaps the most significant factor in our industry: the Baby Boomers are coming. The first ones will turn 80 in 2026 and the needs of this cohort will be enormous. This will result in radical change to the senior living industry, providing lots of opportunity, though plenty of challenges as well. Through my own personal survey of industry peers, the only guarantee is that no one knows what that change will look like. In this uncertainty is found the opportunity for aspiring leaders to be part of an industry that will be re-inventing itself for at least the next twenty years. Such leaders ought to appreciate, however, that they will also be dealing with massive challenges, including how to solve the anticipated shortage of care workers.

Unfortunately for those of us running senior living organizations, there is not a specific date on which the industry will suddenly be transformed; in fact, it is already changing. In order to try to gauge that change, it is helpful for our team members, especially those that are not working on the front lines every day, to engage with residents often, and to *listen* to them. Residents will tell you what they want—the activities they like, the support they need, the technologies they can use. For example, residents can be described as either AC's and BC's; that is, after and before computers, in terms of their comfort with tech. But even that perception is changing, as technology becomes increasingly user-friendly and prevalent. Bill Gates says that we overestimate the change that will happen in the next two years while we underestimate the change coming in the next 10 years. That expresses my point of view. As a result, we pay a lot of attention to innovation, integrating it with resident input to inform our overall strategy. Indeed, since 2009 we have operated a dedicated Center for Innovation and Wellbeing (https://fpciw.org/) to stay on and ahead of trends and opportunities in this domain.

Our overall strategy is guided by the imperative to remain relevant, to both current and prospective residents, including those we will serve ten years hence. We recognize that economics is a serious threat to our relevance. The industry business model is in jeopardy because of rapidly increasing costs, especially labor costs, which are going up faster than the asset growth of our residents. Unless we can introduce new approaches, we will inevitably serve a shrinking percentage of potential residents, because the aging population will continue to grow. Moreover, we are living longer than we expected and longer than we financially planned for, which also places affordability at risk.

The industry's business model is likely to change substantially in the future. Capital investment now is allocated almost exclusively to building and operating quite expensive communities. The future will likely bring more models for congregate living and shared spaces, as well as smaller homes. More people will age in place, unless and until they have some place to go that is both affordable *and* more attractive.

However, in my view, aging in place is not the Holy Grail that will solve the affordability problem. As we age, we increasingly need help, and home care and home health care are expensive. Moreover, as the aging pyramid becomes more rectangular, we will have fewer family caregivers available. Therefore, there will always be a role for the senior living industry, though it will probably evolve to be relevant later and later in peoples' lives.

In my view of the big picture, I see all senior living organizations as interlinked. There is an ongoing "debate" in the industry: which is better, private pay or not-for-profit? But the reality is that for-profits provide superb, cost-effective delivery of the same types of products and services that we nonprofits do. We should stop looking at each other as competitors but rather look at how they do business and learn from them. Certainly, we have done that with fellow nonprofits that we might have considered competitors. Instead, I prefer to view them all as potential partners, whether in operations of a particular location or in sharing a program or perhaps in an even broader partnership. We all can't afford to try to be independent kingdoms. I think it's imperative that we work together, not just for our own organizations' viability, but for the sake of the residents, because ultimately, it's the same service whether, say, Silverado provides it, or we provide it.

For both non-profits and for-profits, the bottom line is economic value and viability. We cannot say, "This program is necessary" and do it without considering what the economic impact is on our residents. I have had so many conversations with residents that start out with them being very angry about some aspect of our business. The more I talk to them, the more I learn that what they are really anxious about is not so much this new service or this small cost that is being imposed on them but running out of their assets. Their conclusion is, to justify this program or that new building, we will throw them out on the street. No matter how much you reassure them, the current generation of senior residents fears being evicted or sent away, and it's heartbreaking. So, what does that mean? First, it means we need to support those people regardless. But it also means that if we are not wise with our use of our own resources, our organization will not be around in the future to provide *any* support.

Aspiring leaders ought to look at senior living as a veritable blank slate. Keep in mind that what we are doing today will have no relationship to what we are doing in 10 years. That should really be very exciting for students as well as those early in their careers. Aspiring leaders should enter the industry and learn the existing systems and flow of things, but simultaneously be looking to disrupt it all. They should be constantly thinking, how should this change, what is a new model that will be better suited for our clients in the future? Anytime there is change coming, there is huge opportunity as well.

Unequivocally, I recommend that people get into this industry. The demand will be huge. Current needs cannot be met. The current model is becoming archaic, obsolete. The scary yet exciting truth is we can make millions of dollars of capital investment now and have no assurance that we will be responsive to the needs 10 years hence. That risk profile means there are great opportunities, and if this is your calling you are in luck.

Doing town halls with residents is my favorite part of this job. The residents can get a little prickly, pushing back and challenging us on our practices, our decisions. This is understandable; such matters do affect their lives, after all. But ultimately these town halls go pretty darn well and our residents are cooperative and grateful and classy. Recently a fellow shouted, what are *you* going to be like when you're in our position? And, effectively speaking for the coming generation of Boomers, I smiled and said, not nearly as nice and polite as you guys! And it is undeniable: we are going to be a pain. After all, in the past fifty years, what institution haven't we challenged?

END

John Woodward is the CEO of Front Porch Communities and Services, a not-for-profit organization with annual revenue in excess of $200 million, that owns and operates both multi-level and continuing care retirement communities as well as affordable housing communities. Front Porch currently serves over 5,000 residents at 38 locations in four states. Prior to joining Front Porch, Mr. Woodward was an investment banker, serving a wide range of not-for-profit organizations, including health care systems, academic medical centers, and senior living organizations. His role involved planning and implementing major expansion programs, reorganizations, mergers and the acquisition of facilities. As an advisor to senior management and boards of directors, he has participated in strategic planning as well as legislative initiatives. He holds a BA in economics and mathematics from the University of California at Santa Cruz and an M.B.A. from University of Chicago with a specialization in hospital administration.

Chapter 10
Pioneering Memory Care with a Culture of Love Greater than Fear

Loren Shook

My dad raised chickens. Caring for them is how I found my work ethic. In kindergarten, I had 500 chickens to feed; in 1st grade, 3000; by 2nd grade, 6000. Within a few years, my brother and I were feeding 100,000 chickens when we were not in school. We worked seven days a week, twelve to fourteen hours a day, even after church on Sundays.

We lived on the grounds of a progressive psychiatric hospital that was run by my aunt and uncle, and I started working there when I was fifteen, doing everything a non-licensed person could, such as dishwashing, laundry, mopping, and even bathing dementia residents. It was grueling work, but since it was only eight hours per day, five days a week, relative to all the years of feeding the chickens, it felt like a vacation. Though because we lived on the grounds, I was on call 24/7 until I graduated from college when I was 22.

The work was deeply gratifying, and by eleventh grade I had decided to follow in the footsteps of my aunt and uncle. Always goal-oriented, that decision led me to study business administration as an undergraduate. After college, I joined a company that specialized in psychiatric hospital administration and soon advanced to a position running a number of our more challenging facilities. Then I started and ran our construction and development division, ultimately becoming President and chief operating officer of the company, overseeing 50 psychiatric hospitals.

After 20 years I left the company and founded Silverado, based on my vision for creating a niche within assisted living: serving memory care patient residents, exclusively. The spark for this idea came from touring some of the typical facilities available at the time and thinking, "I would never let my mother live in one of these places." After rigorously exploring what was considered the state-of-the-art in the field, I conceived a social model of living with proper medical care seamlessly

L. Shook (✉)
Silverado, Orange County, CA, USA
e-mail: lbshook@silverado.com

M. Lifschultz, E. L. Schneider (eds.), *Paths to Leadership in the Senior Living Industry*, International Perspectives on Aging 29,
https://doi.org/10.1007/978-3-030-53966-5_10

layered into it. That included intensive nursing for the residents, and social work services to handle the long-term death-and-dying process for them and their families. The prevailing attitude at the time was, the memory-impaired are lost souls, living without joy or meaning, just waiting for the clock to run out on their lives, and if you merely kept them warm, clean, and comfortable, as a service provider you were doing your job. I thought, that is just so terrible and demeaning. And unacceptable.

By developing a purpose-built concept based, above all, on achieving great *quality of life*, I knew that we could far exceed the appallingly low standards that the industry was setting for itself. The industry was fixated on the costs of owning and operating the real estate, regardless of the impact on residents and their families. My co-founder Steve Winter and I were set on providing a true *living environment*—we would welcome our residents' pets, plant sumptuous gardens, provide intergenerational programs and even provide our own Silverado pets for all residents to enjoy. We would do everything we could to give our residents purpose and meaning in their lives again, because that lifestyle is what I saw growing up on the grounds of my aunt and uncle's hospital. We put everything on the line to pursue our vision—our assets, reputations, really our futures—and started a company that everyone said would fail because it would be so different. We adopted a "Blue Ocean" strategy, which means we intended to create a new market, without competitors. We would recast our would-be competition as referral partners, never considering their products or pricing. Just as Starbucks created a whole new concept of coffee, we were re-imagining what memory care could be.

When we opened our first facility, in Southern California, we actually were offering residents and their families a much lesser physical plant in terms of aesthetics and location when compared to others in the market. People were coming to Silverado primarily from the best retirement communities (CCRC's) in the San Diego area, and they were giving up their private units to move into our companion rooms. Moreover, our prices were higher. Yet not only did we face little resistance, we readily found a receptive market. In fact, over 80% of the move-ins to our 104-bed community came from about 35 miles away, with some moving in from out of state. By contrast, the typical market for an assisted living community is within a 3- to 5-mile radius of the property. We received this response because Silverado's mission was not derived from the grandness of its buildings or locations, but rather a promise we were making to families: *we can change your parent's life*. Implicitly and explicitly we were saying, if the people at Silverado can get your mom or dad re-engaged in life, talking and laughing and finding joy again, how far would you drive, and what would you pay, for *that*?

As we were getting Silverado established and identifying partners to work with, from employees to investors to attorneys to university research centers to public relations firms—that is, anybody who was to be formally associated with us—we were sure to confirm that they knew of and supported the vision and the mission of Silverado. The vision was, providing *quality of life*; the purpose, to change the world. And this has remained so ever since, including when we added hospice and home care some years later.

A good illustration of the risks we took, and continue to take, in pursuit of our deep commitment to quality of life is, we do not restrain our residents. While this was especially controversial early on in our history, it is still a somewhat unusual practice. But because we as founders came out of the healthcare field, we knew the damage being done, physically and emotionally, to people who are restrained. Yes, an agitated older adult can fall and break a hip, but we have always preferred to emphasize the positive considerations of the policy—that we will actually have fewer falls, and happier residents, if we condition and strengthen them and reduce or eliminate the often debilitating medications they should not be on anyway.

This alignment of mission, or purpose, and vision is central to my approach to leadership. And of course, one must start by having crystal clarity on what those elements are and communicating them to audiences in a compelling way that energizes people whom you affect, even including those you neither serve nor employ. By prioritizing compelling communications, you foster a storytelling culture that is another critical point of leadership. Ours, for one example, resulted in a partnership with Fannie Mae, the mortgage giant, which, without our solicitation, approached us to finance memory care for the first time in its history. Their people had heard stories of how Silverado was changing people's lives, and they came to us and offered to finance several communities with $30–40 million. Similarly, we have had research partnerships with such prestigious universities as Stanford, USC, UCLA, and UCSD. What prompted each of these relationships is our reputation for improving lives, which is disseminated through our storytelling.

Effective storytelling and purpose are inextricably linked. You will not convince anyone of anything if you do not believe it yourself. When I was considering starting Silverado, I got the strong sense that God wanted me to be working with dementia individuals, and for me as a Christian that was fundamentally important. When we follow our passion and purpose in what we do, things generally will work out better for us, in part because with that commitment one can find the strength to handle the vicissitudes that we all face in our work. Many factors could have derailed our company over the years, but having that purpose has enabled me, and others, to get through very difficult times.

"Love is greater than fear" is our core operating philosophy; it is foundational to Silverado. The development of this credo resulted, frankly, from experiences mostly in the valleys of my life, specifically from a particularly daunting time when I was facing my second divorce. Though my five kids were doing very well, and Silverado was doing quite well, it was a time of personal crisis. And the philosophy became a source of strength and inspiration for me, and ultimately for us collectively as a team. Indeed, our internal culture audits always identify "love is greater than fear" as our people's favorite quality about the company.

While I can accept credit for operationalizing this philosophy with my team of leaders, I would hasten to acknowledge the great impact that both my father and my coach, Vance Caesar, had in preparing me to embrace it, whether directly or indirectly. Love is a spiritual matter, and I have been fortunate to have some superb spiritual mentors, including these two men.

Love represents courage, kindness, compassion, responsibility, understanding, forgiveness, and all things good. Fear represents the opposite of all of this. We strive always to come from love, basing our decisions and actions on all that love is. In so doing, the positive energy generated by love allows us to offer our residents, clients, patients, families, and associates the most life has to offer and in turn to receive love. Each and every action we take from love serves to build and support the divine human spirit in ourselves and in those whose lives we touch.

Love is what propels our associates to seek innovative solutions to challenges and create opportunities from problems. Our associates are united behind Silverado's goal of giving LIFE—the acronym for love, innovation, family, engagement—to everyone we serve.

Both despite *and* because of the inherent power of "love is greater than fear," it was not easy to introduce and socialize it in a corporate setting. The word love is so evocative that it can be problematic. Indeed, it took my own arduous struggle to gather the courage to bring the phrase to my executive team. When I finally expressed it to them, by way of my personal tale of struggle and learning, they effectively said, "Great, but maybe there's a better word than love." So, we went to a whiteboard, and opened the floor for contribution and debate. Eventually, the idea was settled on to replace love with "like." But when you compare those words side-by-side, the superior power of love becomes undeniable.

We did not and could not find a better word. "The issue is," I ventured, "we are all afraid of the word love. Fear is our problem." So, as a group we surmounted the reluctance, and embraced the word, swiftly rolling it out company wide. Within a few months, just about everyone at Silverado could teach why love is greater than fear. And that became a success that defines our company to this very day.

In a culture audit, interviews revealed that our people defined love in many different ways, so we actually assembled a taskforce to identify the words that we could all embrace as being foundational to love. And that yielded consensus around Trust, Respect, Accountability, Courage, Transparency, Appreciation, Financial responsibility and Purpose-driven. Then we developed a training process dedicated to it. When we are recruiting and hiring, a potential employee must embrace this core operating philosophy because it is embedded in our culture. It is so very powerful that it cannot be faked. And while we have seen it cause some people to leave the company, it certainly has attracted far more to join it.

Leading from love takes courage, and some getting used to, because it does mean that you might have to discharge employees who, in the *short* term, seem to be contributing to the company's bottom line. But that same courage has benefits that far outweigh such considerations. For one, you find that other business comes in, simply because of the power and appeal of love, especially in our industry. Moreover, with "love is greater than fear," I believe you inspire people to work harder because it puts work in such a positive light. When you have a highly motivated, deeply committed team, you benefit by purging the company of the occasional associate or executive who was driven primarily by their fear and their refusal to change.

The COVID-19 pandemic will forever change the way we view the world, how we interact with each other, and how we conduct our business. The implications are vast, and the journey to an end of it will be a long one. For Silverado, how we manage and emerge from this—either stronger, weaker, or not at all—will largely depend on our foundational values and purpose; our "Why." That is, the pandemic will test the strength of our culture's fundamental underpinnings. As just discussed, our culture is rooted in love, and core to love is trust, which in this instance comes in many forms: trust of your leaders in each other and their line staff, and vice versa; trust the families have in the leaders and staff in each Silverado community where their loved ones live; trust in our hospice teams; trust of the regulatory authorities to help guide how we operate as a company in this time.

Fear abounds throughout society, our families, and staff. Political differences divide our country, which foments needless fear, and such fear makes managing through the pandemic more difficult. All leaders must be aware of and deal with it. As it is, the world faces a disease process that is being discovered and adapted to in real time with differing expert opinions as to what is right, a challenge that may be compared to trying to identify and fix the problems of a failing airplane in flight. That is hard enough. Having managed a national behavioral health hospital and healthcare company when the lethal HIV epidemic arrived, I had experience with the medical/scientific world facing a new deadly disease, and I believe that a common thread of COVID-19 and HIV is fear of the unknown. In addition, with the coronavirus there have been massive, unexpected complications, which aggravate the fear. Some PPE suppliers disappeared; others were overwhelmed and out of stock. As a result, those basic but critical supplies were not available from our traditional suppliers for *months* Orders to shelter in place indefinitely, so alien to our nature, caused economic chaos that placed the livelihood and security of millions of people at risk. This caused its own enormous fear. The social and health implications, which may not be fully understood for many years to come, only added to the looming sense of fear.

But at Silverado, we live by our credo that "love is greater than fear." Love and trust are antidotes to fear. Trust is essential in times of great stress. That means trusting others, and others trusting you. Trust allows you to work faster and with less waste and misunderstanding. The "speed of trust" is real, and in a pandemic, it is crucial to have trust throughout the company, at every level, from the frontlines to the boardroom, from the residents to their families. Trust allows you to be nimble and responsive. When we ceased family visitation, we needed the trust of families like never before. Many family members already suffer from guilt and grief from making the necessary but difficult decision to place their loved one in our shared care. Now they were entirely reliant on us to provide that care. We made the option of moving their loved ones out easier than ever and explained that while we were honored to serve them, we would make the transition as seamless as possible. Of over a thousand residents across 19 communities, just a few were moved out; almost everyone chose to keep their loved ones at Silverado. The appreciation and love toward our staff was overwhelming. This was also an expression of trust.

Courage works hand-in-hand with trust, and it too is required in this uncertain era marked by a deadly disease whose behavior is unknown; indeed, our understanding of it changed literally daily. With so many questions to address, early on we instituted a COVID-19 task force comprised of managers and leaders throughout the organization to hold daily calls to review the latest information and make adjustments to our policies and procedures as things develop. We elicit open, transparent, creative, and robust communications from all participants, and through these courageous discussions, better decisions arise. Despite our greatest efforts, some residents have gotten, and may still get, sick and we have to care for them as best we can, even without all the answers. And, the fact is, acting in the absence of complete information also requires courage and trust. Moreover, for our staff who have gotten, or may yet, get sick, we have extended generous sick leave so people can recover without missing paychecks. We have also initiated mass testing of all residents and staff on multiple occasions, which itself has implications that evoke fear from some resident families and staff alike. As always, we work through these fears with love, trust, and courage.

I am proud and gratified that our staff have praised the organization's support, transparency, honesty, and courage. This is a true testament to the Silverado's "Love is greater than fear" culture in action, as well as our leadership's integrity and effectiveness. To be sure, many companies operate with high trust. Unfortunately, though, it is not the norm. When an organization does have such a culture, it will have a significant advantage when facing future adversity. And if there is anything we can be sure of, it is that there will be more adversity coming our way.

Our teams have faced many perplexing challenges. For example, what do you do with a memory-impaired person who is exhibiting symptoms of the disease? Early on, we developed what we believe was an industry first, an observation area that consisted of a separately secured area that had a few units for residents, a common space, and, in most cases, access to an outdoor area. A fire captain complimented the ingenuity of these separate areas because he saw dementia residents locked in their rooms at other facilities. Under extreme pressure, we were able to create these areas across 19 locations nationwide in about one week, which is a testament to the organization's dedication to our residents.

Personal Protective Equipment (PPE) was critically important and, for various reasons, the entire industry's traditional suppliers went to zero fulfillment of orders overnight. We learned that about 50% of the world supply of PPE (such as KN95 masks) came from Wuhan, China, the epicenter of the coronavirus, and their factories shut down, exposing a serious shortcoming to a supply chain that many had taken for granted. We had to find alternative sources quickly. This was not just a problem for us, but for the healthcare industry nationwide. Leadership was tested, as bold choices were necessary to purchase equipment in a marketplace that was poisoned by scammers selling bogus materials and others trying to sell the PPE at upwards of 10 times the normal price. However, with the choice of expensive but indispensable PPE or none, we had to be decisive; lives were at stake all around. Our team pulled together, cast aside the conventions of purchasing, and aggressively sought deals based on calculated chances. By hard work and the grace of God, we were not taken advantage of and we were able to acquire the needed PPE at a mere 2 to 3 times market price. Some healthcare organizations were not as fortunate and went wanting.

The pandemic continues to demand the commitment of the entire organization, from the heroes on the front lines of our communities to hospice staff to the home office. Saving lives entails being on call 24/7. This also requires a commitment from the families of the staff, as our work often interferes with family time and personal lives. The pressure is only exacerbated by the fear of COVID-19, children being out of school, and other realities so many of us must face. Again, I salute our strong culture of love for enabling us to do mission-critical work under unprecedented stress.

To add to the challenge, the media is ripe with sensationalist and misleading COVID-19 stories, often published even as they are refuted with clear documentation. One is shocked to see gratuitous stories whose purpose seems to be to damage a company or even our whole industry. Of course, a byproduct of such media can be damage to the morale of staff, just as they are showing up to care for the vulnerable elderly in the face of danger to their own well-being, as well as anxiety for the families. In a crisis, leaders must deal with such unfair bias through counterprogramming—clear, frequent, factual, transparent, and supportive communications to these stakeholders. You must work to keep staff spirits up and manage the families' fears and expectations, which includes enabling virtual contact with their loved ones when in-person visits are prohibited. We do our best with Zoom, Facetime, and Facebook to show that life in the community safely goes on with laughter and joy.

Yet again, our foundational purpose and very strong culture of love, trust, courage, and transparency rise to the occasion. We see our leaders, under siege from the media and other public institutions, digging deep in their souls to take care of their line workers, showering them with love and kindness. They reach out to the families for support and, overwhelmingly, the families respond by showering them with love and accolades. We get YouTube videos of thanks, and find cards, flowers, and treats left at the doors of our communities.

The pandemic poses a bottom-line business challenge, too. How can leaders manage costs while still meeting the care needs of the residents and hospice patients, as well supporting the staff. PPE expenses run in the seven figures and show no signs of abating. In outbreak communities, budgets break due to staff premium pay, overtime, and recruitment costs for new staff. There are added security expenses for managing the fall-out of biased news articles. There are no easy answers, but a leader must be strategic in financial management in order to maintain a viable organization that can deliver a high level of care. And ultimately, it will be our purpose and culture that gets us through the pandemic successfully.

While our culture is ideally suited to the business we are in, I do not believe that the importance or the nature of mission, vision, and values are different for leaders from one company, or one industry, to the next. These matters are fundamental to good leadership, and good leadership starts with knowing yourself—your purpose and passion, your strengths and weaknesses.

Working with Vance Caesar, I used the book *Strength Finder*, and determined that ideation is one of my greatest strengths. This means seeing disparate facts and factors and drawing them together into a picture that other people do not ordinarily see. One example of this is the concept of "dual diagnoses," or treating people who are both chemically dependent and psychiatrically ill. It was highly controversial when we started looking at the interplay of lithium and 12-step programs, but we had great success with it. As a result of this talent, sometimes my ideas are initially called "crazy," but knowing yourself and what you do well will help you believe in your ideas and fend off critics.

This notion of playing to your strengths cannot be overemphasized. Doing mainly what you are good at is mission-critical for leaders, who generally do not have the time to waste on futile activities. In addition, because leaders are always setting an example, whether they mean to or not, it can be corrosive to your followship to do work that does not result in excellence. I am often reminded of how my mother instilled in me the grit to persevere, to do things even if I was struggling, which generally is a fine quality. But as a leader it is far more productive to surround myself with people who excel at the things I dislike doing or that I don't do well and delegate to them those tasks and responsibilities.

Examining formative experiences can be very useful to developing your self-awareness, which is essential to leadership. When I lived and worked on a farm, many chores were a matter of life and death; being irresponsible was just not an option. I have carried that with me ever since. As a teen working at the hospital, when I was mopping, or taking out the trash, or landscaping, I would always have a patient assigned to me, because my aunt and uncle were committed to giving the patients a purpose, too. And that might have been be a suicidal young person or an older adult with dementia. To many, that would have looked like a risky situation to be in. But I processed these experiences differently: that sometimes you must be willing to take risks in order for people to have a better quality of life. And that became part of our practice at Silverado, including our policy against restraining residents. Especially early on, we had many critics tell us that our non-restraint policy was wrong, that we will be sued. Families threatened to remove their loved ones if we would not restrain. We would explain patiently our "whys" for this practice, and stuck with our beliefs, even offering to help dissatisfied families find another place for their loved ones to live. My sense of responsibility and principled risk-taking are both legacies of relatively early experiences, which continue to inform my leadership to this day.

Responsibility and risk are themes, or factors, that are ever-present in memory care. And this applies as much to those caring for a loved one at home. It is demanding and challenging work, and frequently results in frustration, helplessness, burnout, and, most seriously, *fright*, due to the dangers posed by dementia behaviors, including agitation and wondering. We also see many family caretakers distraught by guilt as they confront the idea of turning over most of the care of a loved one to professionals.

However, this situation will evolve and improve as the Baby Boomers age. Boomers will be more accepting of the services we deliver than their parents were, both in senior housing generally and memory care specifically. That is because so many of us have seen our parents or relatives resist the senior housing option and barely *exist,* then relent, move into a community, and find a much better life there. The social life that is enabled by senior living communities is the greatest enhancement, by far. The isolation of "aging in place," despite its growing popularity as an aspiration among older adults, is, objectively, unhealthy.

The Baby Boom generation also will be more demanding of the senior living industry. Boomers will not want to live as if they are on a cruise ship, spending time only in recreational activities, like the clichéd image of seniors playing shuffleboard. Instead, they will want purpose and meaning and engagement, This will not require a strategic shift for Silverado, however, because we have long emphasized and facilitated meaningful engagement for residents in community activities, whether it is feeding the homeless or training dogs at local shelters.

The dynamic around the therapeutic and treatment programs deployed by senior living providers will also change, as there will be greater scrutiny and discernment of them. That is to say, more rigor will be demanded of us providers, and that, I believe, will necessitate a strategic shift for many operators. Like all change, that will require leadership, both at the level of individual companies and the overall industry. For example, Silverado's Nexus program demonstrates evidence-based results, with enhanced cognition and ease with activities of daily living (ADLs). For Nexus, we developed six pillars of activities—physical exercise, stress reduction, cognitive exercise, specialized digital programs, purposeful social activities, and support groups—and there is evidence that such progressive programs not only improve quality of life but may even slow the progression of dementia diseases.

Indeed, change will become the hallmark of the senior living industry. Technology will play an enormous role in that change. To date, a lot of energy and effort has been put into technologies that really have not yet borne much fruit. Of necessity and opportunity, though, the industry is on the cusp of technological change that will make delivery of care better, easier, and cheaper. The needs are great. For example, facial recognition technology that indicates pain in memory-impaired people who cannot express it themselves is desperately needed, and, I believe, viable. Pain management is one of the keys to reducing bad behaviors, yet too many operators either do not consider this fact or try to manage it with ineffective protocols. Yet when pain is identified and properly treated, many undesirable behaviors among the dementia population dissipate or disappear, and with that, resident patients become calmer and happier.

At Silverado, we have always operated on the premise that people with dementia can do anything they want, such as have pets, enjoy overnight stays with their family, and travel. I have made it a point of my leadership to focus less on risk management and more on the freedom to live life. I believe this has already improved the memory care industry, and as the coming Boomers widely demand that freedom, the industry will continue to evolve for the better.

Another area ripe for big change is how senior living is financed. I have been, and will continue to be, advocating for new thinking around financial accessibility. As we all know, private pay senior living is available to too small a slice of the overall population. Portability in Medicaid dollars is one promising change, whereby governments could permit these funds to be applied more broadly toward assisted living and memory care. However, part of leadership is recognizing that you do not have all the answers, and as of today, neither I, nor the industry, can solve the issue of senior living affordability for the massive middle class that increasingly will want access to it.

The case today, at Silverado as well as in the senior living industry widely, is that we cannot deliver a high level of care for less than we currently charge because the greater cost is in people—the staff—and that is resistant to savings and efficiencies. Innovation is being looked upon by the industry to solve this problem, and there may well be innovations in design, technology, financial instruments, government policy, and other factors that reduce overall costs. I have faith in innovation because it has served me in changing how memory care is delivered.

Another challenge for this industry is understanding how to create an environment of residential services with healthcare components while avoiding the look or function of a nursing home. Many senior living professionals come from the healthcare world, such as skilled nursing, and bring that experience to bear, without recognizing the advantages that assisted living offers. So, the healthcare mentality, despite best intentions, can prevent the full potential of senior living from being realized. Similarly, when people within the assisted living industry move from one company to the other, they often tend to bring the same practices and strategies from these previous experiences and fail to recognize the distinct benefits that the new company offers. People often just want to do the same work they did before, because it is more challenging and difficult to learn new and different approaches. But this business is so personal that employees must assimilate and internalize the culture of each place they work. Ultimately, however, this is a matter of leadership, which must be exerted to ensure consistency and excellence that is aligned with an organization's mission, vision, and values.

Leadership in senior living also requires, in my strong opinion, looking at the "big picture" beyond your own residents, staff, and company, to the industry and even societal level, such as through associations and lobbying groups that can help bring about improvements to the marketplace or the regulatory environment. Historically, the industry has benefited from the sharing of information, such as best practices, even among competitors, as well as presenting a somewhat "united front," whether to regulators or Wall Street, through groups like NIC and Argentum. The willingness to share that "risk," in the form of fees or cooperation, is sometimes foolishly rejected even by those who would benefit from it.

A go-it-alone mentality can manifest in a very different and even more destructive way in terms of decision-making by leaders: micromanagement. The paradox is, the higher up you go the "chain of command," or the realm of responsibility, the less you have to do except delegate to and trust your team more and more. This, of course, can be quite difficult for strong, driven leaders, but the fact is, micromanagement does not work, especially in a personal service industry where companies often operate multiple sites. As with so many matters, this goes back to having a great culture, hiring and training right, and, for leaders, continually getting feedback—from your team, your board, your peers. Good leadership is an ongoing process.

Perhaps the greatest resource for those of us seeking to achieve and improve as a leader is a professional coach. As previously mentioned, I have had one, Vance

Caesar, since we founded Silverado. Our work together has been invaluable to me. Of course, not everyone can afford this service, but fortunately there are excellent options that cost nothing. First, getting a mentor (or mentors) is exceedingly helpful in terms of learning, getting advice, and even opening career opportunities. Second, joining or starting a peer group, especially if comprised of professionals from various backgrounds, can lead to very helpful cross-pollination of ideas, accountability to goals, and moral support. Finally, there is the option volunteering on charity boards, which is not only rewarding in itself and a good civic act for anyone, but can also be a great resource for professional development, as you create relationships with fellow board members. If one were to boil down the requirements for leadership development to a single point, it could go like this: do whatever it takes to learn about and understand yourself, to develop your emotional intelligence quotient.

Listening to *anyone and everyone*, from my board and executive team to our most junior employees, helps me constantly learn about myself and grow as a leader, too. I get open and courageous feedback from our people as to how I can do a better job of leading them, and while it sometimes is unpleasant to hear, it is always worthwhile. So, I must recommend to aspiring leaders as well as incumbent ones: find ways to receive the gift of frank assessment.

Finally, it is critical to leadership development to realize that we learn more from failure than from success. One wishes it were not the case, since there can be great pain in failing. But with that pain comes reflection—about what went wrong and what could have gone better. Contrarily, with success there tends to be significantly less examination. The natural psychodynamics of loss create an openness to change. In contrast, with wins the reaction tends to be "why fix what's not broken?" At Silverado, we deliberately examine even what is not broken, because what is working now may not work later.

In addition, one becomes more self-aware, more prepared, and more resilient by confronting failure. One becomes more receptive to learning and changing. Like a boxer who has been knocked down, you don't just get up and walk into another punch; you keep your gloves up and figure out how to stay in the ring for another shot at victory.

I always get back up. My parents taught me, you are not quitting a sport or neglecting a chore just because things aren't going as well as you would like; instead, you are going to practice harder or work harder, do your very best, learn from your troubles and mistakes, and be better for it.

END

Loren Shook As founder, president, chairman and CEO of Silverado, Loren Shook is often recognized as a memory care futurist. Loren is a true visionary—helping to transform an industry while creating thousands of healthcare jobs—but he is most driven by a lifelong commitment to changing lives.

Loren assures that Silverado has the organizational structure and culture that empowers all leaders to lead, and he enables frontline associates to care for the loved ones of the families they serve. His role is to ideate and lead strategic planning, identify new opportunities, partner with key public policy influencers and further the industry—making it possible for Silverado to achieve its mission and vision to change the world. A leader in senior housing, he frequently speaks at national and international events, and has presented to the Senate Special Committee on Aging—bringing insight on how to provide quality of life for people with Alzheimer's disease.

Loren is the recipient of awards that include the University of California Irvine's Institute for Memory Impairment and Neurological Disorders' (UCI MIND) 2015 Community Leadership Honoree Award, Southern California 55+ Housing Council's Person of the Year Award, Boomer Business Summit's Innovation in Leadership and Service Award, Mission Hospital Foundation's Vision Award, Ernest & Young's Entrepreneur of the Year award in health care and others. He co-authored New Possibilities in Memory Care: The Silverado Story and The Silverado Story: A Memory Care Culture Where Love is Greater than Fear. Prior to founding Silverado, he served as President and COO of Community Psychiatric Centers, an international NYSE company.

Chapter 11
Leadership in Senior Living and Care: Details Matter

Molly Forrest

Writing about leadership is an odd experience for me. I have always assumed that I have led in every position I have held. But I have not thought about how I became a leader. It has been an exciting journey of challenges, opportunities, and a drive to succeed in senior living. At times, being female was a disadvantage in my pursuit of advancement. But in spite of that I persevered and, yes, succeeded in reaching my goals.

When considering a career, originally, I wanted to be a college professor of English literature and to write poetry. But at my first job I fell in love with senior living and from that time, I knew that someday I wanted to be in charge, running my own program in my own way. My particular ambition was to be CEO of a for-profit company in the senior care and living industry. I never considered working for a nonprofit until I was recruited to do so.

My first boss, Robert Frye, was a retired banker who had bought a nursing home to keep busy. After hiring me, he said I had a gift for the field and offered to train me, though he had never before trained anybody for this work. As an English major, I knew little about business, and less about healthcare policy, Medicare, and Medicaid. In my year of training I worked six days a week, took few vacation days, and, with his encouragement, attended classes in anatomy, physiology, microbiology, medical terminology, accounting, finance, and health care administration. He also required me to learn how every aspect of the operation functioned, including the kitchen, where, as a final learning, I once prepared lunch for the 49 residents. When I ran a nursing home years later, one day the two cooks didn't show up. I hustled into work and cooked three meals for 120 residents. Robert Frye gave me the training, the education, and the confidence to tackle many challenges.

M. Forrest (✉)
Los Angeles Jewish Home, Reseda, CA, USA
e-mail: Molly.Forrest@jha.org

M. Lifschultz, E. L. Schneider (eds.), *Paths to Leadership in the Senior Living Industry*, International Perspectives on Aging 29,
https://doi.org/10.1007/978-3-030-53966-5_11

He was a believer in what some call management by walking around, or what he called making rounds. That meant responding to a call light if somebody wanted to turn off their TV, picking up a piece of trash, or dealing with a wet spot on the floor, whether you put a sign on it and found a housekeeper to clean it up or just grabbed a mop and did it yourself. If I could handle a task, then I did it. "You are part of a team," he said. "While you are the leader, you must role-model the behavior you hope others will follow." He also dispatched me to run two buildings in another city, and advised, "Don't fool yourself to think what you did here will work elsewhere. What you bring from your old place are experience and ideas. But problems and solutions vary wherever you go." I have always kept that wisdom close. It really means, never think you know everything and never stop learning.

I left his employ to move to Oregon with my husband, taking jobs and seizing opportunities as they came. It was a different age. Early in my career, I heard from many for-profit companies and nonprofit organizations, "If you get pregnant, you'll be fired." This was not an uncommon practice at the time. So, I took a job in county government where sick/maternity leave was available. By no longer being on call 24/7, it allowed me to have work I enjoyed *and* to become a mother.

First, I supervised an Older Americans Act county-wide meals program. Then I managed a housing and services program, where I worked with seniors, veterans, abused women and children, the disabled, and other vulnerable populations. When they learned I was a licensed nursing home administrator, I was appointed the county's first Ombudsman, a position with considerable reach but limited authority to address issues for nursing home residents.

In 1979, I moved to Los Angeles. I had been told in Oregon that if you can "make it" in LA you can make it anywhere. It was challenging. As a single mother of a two-year old, I faced discrimination. At the time, employment discrimination laws were not what we have today. Potential employers would actually say, "With a young child, you will not be dependable" and "Corporate leadership is exclusively male and closed to women." Today, discrimination still exists, but it is more covert. When confronted with such challenges, I sought to either navigate around it, blast through it, or turn and take another path.

I meant to spend three years in Los Angeles, gain more experience, and then return to Oregon and accept my parents' offer to buy me a care facility to operate. Fate intervened. I fell in love with a man who was rooted in LA and I accepted *his* offer—to marry. I also fell in love with working in Los Angeles. I felt called by its creativity and diversity, as well as the prospects of helping improve the lives of a large senior population.

I soon took a position with a Midwest-based nursing home operator and was told that if I was successful, corporate management opportunities would be available. The LA facility they had just purchased had been featured on the front page of the *Los Angeles Times* as the "worst nursing home in California." After one year under my leadership, the facility became a training center for the area's Department of Health Licensing and Certification, for exemplifying a well-run nursing home. At the company's annual leadership dinner, I received awards for most improved facility, highest occupancy, and several other key benchmarks for nursing home management, validating my success. To my surprise, that night the company announced the promotion of a male colleague to supervise all of their facilities in California. I was unaware the job had even been open.

I filed a complaint with Human Resources but said, "I won't sue. I just want the chance to apply for the position when he leaves, which is inevitable because he does not have the experience for the job." As I predicted, after a year he left. To find a replacement, the company engaged executive recruiters and advertised the job extensively.

Ultimately, I was hired. Moreover, the company expanded my role, putting several states and more facilities under my direction. My career path there was promising until the firm was bought out by a company whose CEO repeatedly harassed me, despite my strong and consistent objections. When an egregious incident occurred in front of my direct reports, I resigned.

I decided to start my own consulting firm in Oregon, which enabled me also to care for my ailing father and my mother who had developed Alzheimer's. My consulting practice thrived, but the lifestyle was brutal, as I traveled to see clients Sunday-Thursday and had precious little family time. By January 1996, my father and mother had passed, and we moved back to LA.

A friend in the industry asked if I would be interested in interviewing with the Executive Committee of the governing Board of the nonprofit Los Angeles Jewish Home; they were seeking a new CEO. After a three-hour interview with most of the 21-member Executive Committee, I was offered the position. I had gotten what I had long sought: a top corporate leadership role in the field of senior living. Though as I later discovered, I still had much to learn.

Initially, the Home had sought new leadership to manage through its financial straits. For 84 years, the Home had served the "needy." Typically, that meant the financially needy. As a result of this noble mission, in 1996 over 70% of the Home's residents were on Medicaid, meaning they lacked financial resources to pay for their care. Medicaid rates in California support basic subsistence, but not the quality care expected by the Home's constituencies. Because the Home stepped in to fill the financial gaps and provide such care, it was in financial distress. I was confident I could improve the situation. However, I knew very little about nonprofit organizations in general and the Jewish Home in particular. Moreover, the dynamics of Board governance and politics eluded me. The first years were difficult growing years for the Board and me

After my first three years, the Board Executive Review Committee informed me that I had done well fixing the pressing financial issues but that I lacked the "stature" to continue as the CEO of a now financially healthy organization. They offered to demote me to COO when my contract ended, contingent on approval from a new CEO. I was devastated. I promised to raise my public stature in the industry and the community so I would be more visible.

When a year later I presented the articles I had published, the speeches I had made, and the boards I had joined—all to improve my stature—a Board Committee member said, "You're too short to be the CEO. No one knows when you enter a room. Your voice is too quiet." When I shared this with my husband, he said, "What he meant was you're too *female* to be the CEO." Though upset, I resolved to surmount this setback.

I asked the Board to extend my contract for three years, which they did, provided I work with an executive coach. The coach was quite savvy, and he observed that I was trying to persuade the few Board members who opposed my continuation rather than cultivating the many who supported it. He advised me, sensibly, to spend more of my time strengthening these relationships. I took his advice. I have continued as the Home's CEO for the last 24 years.

Leaders must always be aware that while you may be at the top—whether of a team or an organization—you still have to satisfy the various constituencies you serve. In the case of being a CEO, besides your board of directors, you also need to satisfy regulators, the community, your staff, your funders, the residents, and the families of the residents. You may think you're the boss, but in reality, *you* have many bosses. In fact, my work as CEO largely takes the form of satisfying them—often more than one constituency at the same time.

After I became CEO, I realized that fundraising would be a high priority for me. Our board had established the admirable goal of serving all those who need care. As a result of this laudable goal, most of our nursing home residents were (and are) on Medicaid, meaning they do not have the financial resources to pay for any of their care. In terms of government reimbursement rates for such patients, the state of California is in the bottom ten states nationally; it pays slightly higher than half of what New York pays. At the outset, we needed to raise $5 million *annually* to fill the revenue gap created by providing excellent care for our long-term nursing residents. Today, this deficit is approaching $20 million a year. It is instructive that many non-profit nursing homes in California control the number of Medicaid residents by limiting the number of nursing home beds, to avoid the need for such massive fundraising.

While the previous administration of the Los Angeles Jewish Home was focused on serving the needy, the reality of the increasing yearly losses from providing quality care to indigent residents created great financial stress. And, in addition to the need to make up the structural deficit to provide good care to our residents, I inherited two campuses of older, unattractive buildings that would require expensive updating. Finally, we were also planning ambitious new programs, which necessitated additional funding.

So, to continue to provide good care to our residents, to build new buildings, and to start new programs like hospice and PACE (Programs of All-inclusive Care for the Elderly), I would have to raise large amounts of money. Los Angeles does have a large philanthropic community, but there are many organizations competing for their dollars, including USC, UCLA, American Jewish University, Hebrew Union College, Jewish Family Services, The Jewish Federation, and others. Fortunately, after 88 years of serving seniors, the Home had historic connections to thousands who were very pleased with the care the home has given to their loved ones. In the 24 years that I have been the CEO of the home, I have been able to secure the funds to compensate for Medicaid's parsimony, to build several buildings on both of our campuses, to build two state-of-the-art Life Plan communities, and to enable us to launch and maintain 11 new programs such as PACE. In 2014 we set a philanthropic goal of $270 million; to date $220 million has been raised. During my tenure, the four largest donors have been women. It is a nice byproduct of their generosity that their gifts have allowed a woman CEO to succeed.

Fundraising is important, but the greatest responsibility I have is the leadership and management of our staff. Your workers are simply the most critical asset in senior living, as in almost any industry. This is especially the case in an industry that has been and will continue to be bedeviled by staff shortages, low morale, demanding work, low wages, and chronic under-funding by government programs. Looking ahead, it is projected that the senior living and care industry will need in excess of one million additional workers in the next 10 years, yet no one knows from where they will come. Day to day, particularly for females, working as a certified nursing aid or as a nurse in long-term care is considered by OSHA to be one of the most dangerous jobs in the United States. Nurses lift and move the immobile, care for dementia victims, and, as of 2020, face exposure to COVID-19. You are also dealing with people dying, their grieving families, and your own grief, for these are people you cared for and with whom you had relationships. Through all of this, you are expected, if not required, to remain calm and perform stressful work mistake-free. Considering all of these factors, it is clear that leadership is about tending to your people—keeping them safe, satisfied, and motivated.

Since arriving at the Jewish Home, I have always enjoyed meeting with our staff and learning from them. At regular intervals, I obtain their feedback through direct conversations without the participation of supervisors. In each meeting, I begin with an overview, allow questions with open dialogue, and also ask them to complete an anonymous "morale survey." And I *listen. By* doing so, I try to give all of our staff an engaged and relevant voice, and I also, crucially, learn so much about what is happening "on the floor" outside of the executive suite. I also created a form of staff meeting that we call a "Speak-Easy," a play on the speakeasies of Prohibition, which were illegal taverns that offered alcohol, gambling, and prostitution. You could say that our Speak-Easies share only the informality of the earlier kind, as we gather to talk about work and life in a relaxed and open forum. They have been so well-received by the staff, and helpful to me as a way to connect with our 1600 staff members. In twenty annual meetings, I generally share information on the impact on work of events that are happening globally, nationally, statewide, and within the organization. They share with me whatever is on their minds.

I also stay in touch with the mood, preferences, and morale of our staff by reviewing every one of the hundreds of responses from our regular morale surveys. We publish summaries of the issues raised and comments made for all staff to see. I vet the actionable items with our executive staff and make what we can a reality.

I also believe deeply that it is a sacred obligation of leaders to cultivate the next generation of leaders, not only for the sake of the organization but for the sake of those individuals, too. And I try to do so no matter where the staff may be on an organization chart. Here are two examples. One, our current COO came to us from outside the Jewish Home, at the strong recommendation of one of our board members, USC's Dr. Ed Schneider. While she was not as familiar with the culture and operations as someone from within would have been, she is so enormously talented that I have been pleased to devote whatever time necessary to share the Home's ways with her. Two, a recent grad had an internship with us. When we met, I asked her, what are you going to tell people you accomplished here? I wanted to be sure that she would be able to say she did something tangible. She gamely took on the task of rewriting our resident handbook, which was obsolete, and did such an impressive job that we hired her for a full-time position on a two-year leadership training track. More broadly, we invest heavily in opportunities for personal development, such as having nearly 2000 educational online videos available to staff. We also are supportive to the extent possible when staff are interested in attending school or pursuing a degree; we support *learning*.

Language is one of those areas. Many non-English speaking staff work in senior living, including our organization. In fact, in the beginning, the majority of the Speak-Easies had to be translated. I wanted to encourage people to improve their professional prospects, as well as our performance; learning to speak English would help accomplish both. We pay for English workbooks from the local high school. We reward the staff for completing and "passing" each workbook, assisted by a local retired teacher (who volunteers with us). As a result, only 2 of our Speak-Easies still need translations. It is beneficial for our staff and is good for our residents.

Another priority of my leadership is to improve staff performance and job satisfaction, and one way I try to do this is by emphasizing the importance, and joy, of teamwork. I believe strongly in management by objective. I particularly like two types of objectives, which are so much more interesting than conventional ones: one, I want our people to have goals that require them to stretch their abilities; two, I like goals that are focused on team performance. In addition, I promote teamwork and friendship among staff informally. I once had a social worker and a director of nurses who had to work together but were often at odds. Staff conflict can be corrosive, so I intervened. We all met, and I told them that they were great in their jobs, but could be even better if they got along. Since I knew and liked them both, I suggested that they didn't see each other for the unique individuals that they were, and I nudged them to share lunch once a week but *not* talk about work. And, sure enough, they became friends and even better work partners.

In addition to teamwork, two qualities I look for in people are creative thinking and critical thinking. Creative thinking is about new ideas and likely outcomes; critical thinking analyzes existing ideas and finds better solutions. Leaders know that every organization needs both, for we always have to be asking, how well are we doing things now, how can we improve, and what else can we do next. In interviews, I seek to engage both types of thinking on the spot by asking unexpected questions that engage their brains and defy easy stock answers. I happen to love asking, what organizations have you learned the most from, what did you learn, and why was it helpful? This compound question will often yield revealing evidence of one's creative and critical thinking.

Such thinking is important because senior care and living is a complex industry marked by extensive regulations with Byzantine rules, rapid changes, and ever-increasing costs. A skilled nursing facility, for example, has to understand, comply with, and respond to more regulations and on-site surveys than a nuclear power plant. In senior living, we are faced with an underpaid workforce, shortages of such workers, unstable finances, increasing litigation, and rapidly changing demographics in the form of an aging population. What makes the work even more challenging is we are caring for the most vulnerable and dependent members of our society. Thus, success in this business requires leaders to be everything from detail-oriented information processors to big picture problem-solvers. As the leader of our organization, I have to do it all.

Leadership is required to be aware of what's coming in the future and to adapting to the changes in senior living. When I was hospitalized for neck surgery, the Affordable Care Act was being assembled in Congress. I was bedbound for three months before the surgery and three months after. I took the opportunity to read every word, including the commentary, of both the draft of the Act and the final adopted version. It was 13,000 pages long, with tens of thousands of pages of commentaries. Buried within it was a program designed to try to reduce the 33% average hospital re-admission rate within 30 days of Medicare patients having the same condition or an exacerbation of it. I felt like I had found a hidden pot of gold, because I knew we could do very well in such a program and perhaps attract additional funding as a result. We applied to participate. Just 102 providers were selected, including some of the finest hospitals in America, such as Cleveland Clinic, UCLA, and Cedars-Sinai Medical Center. The Los Angeles Jewish Home was selected, too. After three years there were 35 programs left, including ours. And we were recognized as one of the two best providers in the nation. We reduced the hospital readmission rate of the three hospitals that we dealt with from around 33% to 12%. Unfortunately, with political change, the program was cancelled, but I was very proud of our performance. I still believe the program we developed, operated, and proved successful could be meaningful in controlling unnecessary re-hospitalizations of Medicare beneficiaries.

Starting a Program for the All-Inclusive Care for the Elderly, PACE, for the Jewish Home is another example of detail-oriented and global leadership. PACE is often called "a nursing home without walls." The intent of the program is to provide all the care and medical needs for poor frail nursing home-eligible seniors except housing. The federal goal is to provide better care in homes in the community and lower the cost of care to Medicare. Due to the highly successful outcomes of care, California recently opened PACE to for-profit organizations. To me, the entrance of for-profit motivation in the healthcare of the in-home frail is concerning. Their overall condition and care will be less visible than it would be in skilled nursing facilities. Sometimes the profit motive in healthcare services seems to conflict with a stated governmental goal to provide quality care and at lower costs.

Like all state and federal programs, it took a mountain of paperwork—and also $5 million—before the agencies would approve our PACE proposal. We were fortunate to have generous donors, Joyce Brandman and her late husband Saul Brandman, provide the funds needed to start the program. PACE reimburses providers based on an acuity score. That is, the more serious conditions a patient has, the more money the government provides. As a provider, you want acuity scores as high, and of course as accurate, as possible. Initially, however, our organization was not paying attention to our acuity scores, and when the scores dropped, we lost money. We had to find consultants to train our staff and hired a new medical director. Now our scores are higher, and we're making 3–5% "profit," which means we can pay all of our PACE bills.

Even with the best possible acuity scores, running PACE can be very costly for us. For example, if we are being reimbursed $12,000 a year for one PACE enrollee, it may sound like the program is well-funded by the government. However, if that patient is on dialysis three times a week, and he has, for example, Hepatitis C, which requires a drug that costs $7000 a month, we incur large losses on his care. It is a capitated program, which means the federal government and the state of California expect us to pay all his costs.

This is the age of managed care. It is difficult for an organization like the Jewish Home to survive outside a managed care organization. Our residents are bombarded by these organizations to join their plans. Senior living executives cannot be passive in this new world. A leader needs to deeply understand and assess the options carefully. We are in the process of starting our own IPA (Independent Practice Association) with our own in-house physicians so that we can be in the best position to negotiate with large healthcare providers. We are also investigating becoming our own Special Needs Plan, a Medicare Advantage option. It is exceptionally complicated work. I am closely involved in all the planning and execution; I need to understand how it "works" for the Home. I may leave the details to others but overall I must understand enough of the details to assure myself we make a sound decision.

Small things matter in role-modeling behaviors we expect of others. Sometimes "details" might include a piece of trash, which I can spot from a mile away. As Robert Frye taught me, I go and pick it up, and I expect every manager who works here to exhibit the same concern. The fact is, we are here to provide care and a wonderful living environment for our residents. If one happens to drop a dirty tissue on the ground, you must pick it up, dispose of it, wash your hands, and carry on. We have the privilege of working here, but it is our residents' home and we need to treat it like their home. We also have to be frugal. We may not have the best flooring, but we keep it clean. Modeling behavior, being clear about expectations, rewarding and celebrating victories, are very important parts of my leadership. As the saying goes, I have to walk the talk. I model our policies and procedures and look to inspire our staff and volunteers to adopt our mission and values. I must make it a regular practice to communicate our mission, vision, and values to our staff and our residents.

It took us eight years to produce a new mission statement that captures our essence. A succinct mission statement relevant to the times boldly defines an organization and sets a course for the future. Ours is very succinct, but very ambitious: to provide "excellence in senior care reflective of Jewish values." Our core values focus us on Charity, Quality, Dignity, Jewish Values, and Fiscal Responsibility. Our mission statement and core values are not hidden in my desk, but widely promoted in every venue. Everyone knows them. As an organization, we live by these words.

I believe that one of the most common mistakes made by leaders is bad decision-making. Whether it's rushing to judgment without studying all the facts, or the flip-side of being paralyzed by overanalysis, I believe having an operational mission and values statement can improve the decision-making process. Making a decision in and of itself is a process consisting of several steps: gathering the information, organizing it, analyzing it, evaluating the options and potential outcomes, and ultimately making a reasoned judgment that leads to your decision. However, looming over the final decision also is the ethics part; is it the "right thing" to do? Is it the best thing for all concerned? Such questions are often at the heart of my decisions. Mission and values will often help clarify any ethical ambiguities because they dictate the goal for the decisions that you make.

Of course, there are decisions to be made that are not at all ethically ambiguous. We made several relating to COVID-19. Like many senior living facilities, we had to respond rapidly when the infection started to spread.

The first hotspot for COVID-19 in the United States was a nursing home in Kirkland, Washington. Our nursing home residents were particularly vulnerable. With an average age of almost 90, many of these residents have conditions such as diabetes, heart disease, and hypertension that increases their risk of death from infection.

Our executive staff began our preparations at the end of February 2020. Our Pandemic Disaster Plan was adopted, and daily disaster management meetings were initiated March 4, 2020. We immediately restricted outside community events for residents, restricted visitors from coming into our buildings, updated and reinforced infection control/prevention practices, and reviewed our ability to protect staff and residents. The fact that this would cause certain inconveniences and upset daily rhythms and activities did not compromise our decisiveness. Many of the protective measures we took were implemented *before* guidance and requirements came from CMS (Centers for Medicare and Medicaid Services), State or local public health authorities.

Initially, we had no COVID test kits, so we took temperatures of staff and vendors and gave them a questionnaire about symptoms and exposures. Like many other senior living facilities, we did not have enough personal protective equipment, so we modified existing gowns. We designated a part in each of our buildings as a quarantine area.

We initiated daily meetings of our COVID-19 task force, which includes all officers, executive directors, administrators, clinical leaders, and managers, a total of 60–70 staff. We meet 7 days a week. We prepared to take whatever steps necessary going forward to protect all of our residents and staff, particularly those who are most vulnerable. A forecast for the future is this: the national concern for the pandemic's impact on congregate senior living centers will change long-term care as we know it.

Indeed, the future of senior care and living has always had challenges and opportunities for aspiring leaders. COVID is one factor that demands we create new answers. Another is that the Baby Boom generation is here, which means a huge population of people living longer and in need of housing, services, and ways to enjoy life in the older years. This means the next twenty, thirty, forty years will be dynamic ones for this industry. Find your passion in this field, find a mentor who supports your learning, encourage others whom you work with to pursue lifelong learning. Find ways to cast your bread upon the waters.

I work to support my organization today, but also recognize that I must do what I can to advance senior care for tomorrow. I run the Jewish Home, serve as Chair-elect of LeadingAge California, lecture at USC's Davis School of Gerontology, serve on several boards in the Los Angeles community, and, whenever invited, go to Washington, DC and meet with members of Congress and friends at AARP. I'm just over 72 years-old and will shortly commit to a final two years with the Jewish Home. Every day, I feel I am just getting started.

END

Molly Forrest is a senior living industry executive who has dedicated her career to care of the elderly. Since 1996, Ms. Forrest has served as the Chief Executive Officer and President of Los Angeles Jewish Home for the Aging (LAJH), a non-profit serving 4,000 seniors in-residence on four campuses or living in their homes through a wide array of community services. She led the Home through the most ambitious expansion in its history. Today, the L.A. Jewish Home is a diversified healthcare system, offering in-residence settings for independent living, residential and assisted living, skilled nursing, dementia care, rehabilitation services, and an acute psychiatric hospital. Community programs offered to assist seniors in their own homes include Brandman Centers for Senior Care (a PACE organization), Home Health Agency, physician group, medical clinics, Skirball Hospice, and Jewish Home Center for Palliative Medicine. The Home's Annenberg School of Nursing is nationally recognized as a quality educational/clinical program. During her tenure, philanthropic donations to the Home increased from $1.5 million in 1996 to $18.1 million in 2019.

Ms. Forrest's earlier work included leadership roles in senior services, including for-profit organizations and governmental programs serving seniors. Her recognition includes the L.A. Commission on the Status of Women's "Pioneer Woman", L.A. Business Journal's Healthcare Lifetime Achievement and "Women Making a Difference" and McKnight's Women of Distinction 2020 awards. Ms. Forrest is a well-received public speaker for business and association meetings, and the Chair-Elect of the Board for LeadingAge California. Ms. Forrest is married with four children and six grandchildren!

Chapter 12
Bringing a Competitive Spirit to Leadership in Senior Living

William (Bill) Pettit

Dissatisfaction is native to leadership. Leaders live in the gap between what is and what ought to be. Leaders do not experience or tolerate complacency; "good enough" is not in a leader's vocabulary. If a leader's chief role is to create a vision and to motivate others to join in to achieve that vision, then a leader is required to have a competitive spirit.

Competitive sports have played a big role in my development as a leader. I have been a competitive athlete all my life. Whether it has been playing Division I college lacrosse, or hockey, golf, or tennis at the club level, I have learned significant lessons about myself and others from sports; competition is a great teacher. In competition with others, you learn about values and principles, winning and losing, team dynamics, motivation, and work ethic—your own and others. From a relatively young age until today, on the playing field and off, I have had a desire to lead, to optimize performance, and to enjoy the benefits of success, both as an individual and as part of a team or organization.

To some, competition has a bad name because it *ostensibly* produces only winners and losers. As one who is devoted to coaching along young, aspiring leaders, I reject that mentality. The meaning of competition goes well beyond just winning and losing. It actually helps foster emotional intelligence (EQ) and social intelligence (SI), both of which are key to developing relationships. Senior living is an industry where that ability is a *sine qua non.*

My formal education also contributed to my development as a leader. As an undergraduate, I studied psychology, which I chose principally to help me gain a finer appreciation of how people think, what motivates them, and how they deal with challenges. Then I attended graduate business school and received an MBA

W. Pettit (✉)
Merrill Gardens, Kirkland, WA, USA
e-mail: billp@merrillgardens.com

M. Lifschultz, E. L. Schneider (eds.), *Paths to Leadership in the Senior Living Industry*, International Perspectives on Aging 29,
https://doi.org/10.1007/978-3-030-53966-5_12

with an emphasis in finance and real estate, which led directly to my careers in banking and then senior housing.

Together, my training in sports, psychology, and management reinforced my innate sense of how to lead primarily by providing an informed perspective on the people who are part of your team, especially in terms of how to develop relationships with them. This also taught me the importance of what one can call *the law of personal nuance*: not everyone looks at the same things in the same ways. That applied insight translates to empathy—the ability to put yourself in the shoes of others. When you contemplate leadership, vision, and execution, I believe these factors tie nicely together.

After graduate school, I became a financial analyst in commercial banking. This led to a series of challenging and rewarding positions. After heading a bank's strategic planning and acquisitions group during the deregulation of the financial industry, I was made chief financial officer of a publicly traded bank, and ultimately became president of a regional bank. Despite having a successful banking career, I decided, at age 40, to try something entirely new, and I took a job with an old-line, family-run timber company to help them identify and exploit new commercial opportunities in other industries. This could be seen as an example of my credo that healthy dissatisfaction with the status quo is a necessary requirement for a leader.

Despite being a great opportunity, making this move was also a great challenge; indeed, the two often go hand-in-hand. I was relatively young, was well-compensated, and had the respect of being the president of a publicly traded bank. While the road ahead in a dynamic new industry looked exciting, I was not offered much more than a skeletal staff and a small amount of capital to invest. But I had reached a point where my work in banking was not motivating me on a meaningful level. After spending 18 successful years in banking, I had concluded that it was not as fulfilling as I wanted my work to be, chiefly because I had virtually no contact with our customers on a daily basis, and I am a people person. When I was offered the chance to shape greater meaning into my work, I seized the opportunity. All I had to do was explain to my wife why I was taking a 50% pay cut for an uncertain future.

At these tough moments—call them personal inflection points—one is fortunate to be able to draw on the wisdom and inspiration of influential figures in one's life. I had two. My father was a huge influence on me; he instilled in me a sense of ethics, integrity, work ethic, and teamwork, and armed with those qualities, I believed in myself, which made it ok to take a career risk. I also was mentored by Richard Cooley, who became a business legend while running Wells Fargo Bank but eventually left and, improbably, joined the financially troubled bank where I was CFO. I asked Cooley why he left a prestigious institution to join one whose future was in doubt, and he explained that the secret to the longevity of his career, and his enthusiasm for his work, was his practice of reinventing himself every five to seven years, actively seeking a new challenge—something that would stretch him while keeping him engaged and motivated. Having learned profound lessons from these men, I was able to leave my comfort zone for uncertainty.

The first business we started, in 1993, was our senior housing company, Merrill Gardens, and that is where I remain CEO today. Basically, we started with a group of individuals and some property in foreclosure. We wanted a proof of concept, so we demonstrated that we could fix the buildings in foreclosure, *and* that we could build quality assets from the ground up on a multi-state basis. It took about five years to demonstrate that successfully. Today, we have 70 communities in twenty states offering independent living, assisted living, and memory care. And we are growing, with nine new communities in development, plus control or ownership of another six. We also have three communities in China.

Working for a family business has its challenges, but also great benefits, the foremost being the ability to think long term, rather than from one fiscal quarter or one year to the next. While some institutional investors do think in terms of seven to ten years, even *that* is short term compared to how time is viewed in the timber business, which is where the Merrill family is rooted. Typically, the family plans in terms of *generations*. With that long a time horizon, and the freedom from having to report results to Wall Street each quarter, being a company in the senior living industry we enjoy a considerable advantage. A leader can utilize long-term planning and execution to adapt to changing populations and technologies. In addition, I was given a virtually clean slate by the family.

I was able to utilize my experience, hire the right people, and create the right vision in an industry that presented extraordinary opportunity but was really just starting to become professionalized at the time we launched Merrill Gardens. We started with core values, vision, and strategy, as well as a conviction of the general type of company we wanted to be—excellent, innovative, and responsive to our community. Looking back, what I was doing as a new leader was building the vision without the baggage of having to change the culture of an existing organization.

In an industry like ours that struggles with high turnover, every organization needs to put their mission, core values, and vision out in front of every new hire to the organization so that everyone knows what is expected of them. For us, this starts with each and every individual Merrill Gardens employs, from the moment of one's orientation to the work done on a daily basis.

Recruitment of personnel should include ascertaining whether an individual's core values are consistent with the company's, and whether, as you explain the mission and the vision, they feel comfortable not only embracing but being enthusiastic about this. Given the high stakes of our industry and the relatively modest compensation of the jobs industry-wide, it is a leader's responsibility to help create working environments that attract great people.

Another way to make senior living careers stickier is to create a culture that encourages and rewards creative thinking. Our culture at Merrill is designed to tolerate mistakes, and indeed to use mistakes as learning experiences: I call this "try it, fix it, do it." So much of what we do in this industry involves encouraging people to live the mission of the organization, and in doing so, encouraging them to think of better ways to execute that mission and vision. We really are grounded in the belief that the search for the perfect solution is nowhere near as important as moving the organization forward through innovation. And innovation is rarely perfect the first time out. By fostering a cultural vision of experimentation, correction, and execution, I believe you not only move the organization forward, but also give your professionals a deep sense of investment and pride in their work, because you are demonstrating trust in and appreciation of them.

Our field is a pure meritocracy. This is what makes careers in our industry more attractive than many other industries. I teach this, preach this, and discuss it all the time with our young managers, and believe that this is at the heart of why senior housing careers make for extraordinary launching pads. If you are excellent at your work, the sky is the limit. The principal challenge is that senior living poses an extraordinarily complex management and operating model, and there is no way one can receive adequate training for it in a formal setting. No school can teach, and no degree can confer, the actual qualifications required to be a star in this industry. The closest degree would be the new Master's degree in Hospitality and Gerontology offered jointly by the University of Southern California and Cornell University. Such schools can teach the technical elements of customer service, gerontology, and humane design, though not the complexity of operating in a regulated licensed industry. Another emerging program can be found at The Granger Cobb Senior Living Institute at Washington State University. Consequently, our best leaders in the industry to date have been, by definition, largely self-taught. However, this may change with the new programs being offered, coupled with internships in senior living companies such as Merrill Gardens.

There *are* supervisory roles for inchoate professionals, but these aspirants, no matter how innately talented or well-educated, will not have the full complement of skills necessary to *lead*. We seek two kinds of skillsets, two distinct types of managers, who can thrive: people who are entrepreneurial and solution-driven, and those who are process- and value-driven. Sometimes, we do hire on the principle of "best available athlete," meaning we will bring on someone who has both potential and some useful experience in the industry, then train them on the technical aspects and our particular approach. But what cannot be trained or changed is one's aptitude for leadership.

The population is getting older and needs help. At Merrill, for example, we looked at China and profiled the wants, needs, and expectations of elders there and, somewhat unexpectedly, found that they were substantially the same as those of elders here in the U.S. It suggested to me that aging is the same regardless of what nationality you are. Your fears, your concerns, your needs—they do not necessarily change from one country or culture to the next. They just manifest themselves in different ways.

As we developed our business model, we realized that the opportunity to serve elders in China was enormous. Their population over the age of 65 is approximately ten times that of the U.S. There were almost no available alternatives for Chinese seniors and there was a heavy dependence on a culture that emphasizes respect and responsibility for elders on the part of the next generation. But all those factors were in conflict with the realities of the country. The Cultural Revolution and the One Child policy was in direct conflict with this generational sense of responsibility. Most Chinese resided in a village-based, agricultural society. As China industrialized, it focused on creating a less expensive labor pool and adopting technology to create a manufacturing society. Out of necessity, this pulled the younger generations out of those villages, and put them into huge cities and manufacturing complexes where pay and resources provided almost no space or opportunity to care for their parents by bringing them from the villages to the city, effectively leaving seniors behind. This is another example of where challenge and opportunity meet.

In China, the challenge is that while there is an emerging middle class that is quite prosperous, there is almost no wealth in the older population, because they are all products of the original Cultural Revolution. It is the adult children that have the cash, and there are ample alternatives, such as nannies who work at very low wages and live with families, providing care to the elders in the family because the family cannot or does not have the time to care for the elder. They have yet to adjust on a mass scale to the value proposition for professionally run senior housing in China. It is still a number of years from being a preferred solution for elder care on a privately funded basis in China. In addition, unfortunately, the government programs for elder care in China are depressing in terms of environment and service support. So, our challenge was, and continues to be, how do you build a business environment there in terms of price, scale, and services. This is what we have been puzzling over for the last six or seven years. This challenge of developing a business in a rapidly changing environment featuring uncertainty will certainly be duplicated by the coming generation of leaders in senior living, here in the U.S., as well as in Australia, Japan, Canada, the United Kingdom and Northern Europe.

The senior living industry was started in the U.S. primarily for the Greatest Generation, but of course we are rapidly headed for a "collision" with the Baby Boom Generation. One mistake people might make is looking in the rearview mirror at what was successful over the last two or three decades and ignoring the signs that serving this new generation will need to be different. They expect and will require it. When we started in this industry, operator builders looked mainly for cheap land, then typically built buildings with utter indifference to whether they looked and felt like a residence. The prevailing wisdom was that people will enter senior living because they *have* to, not because they *want* to. Every design component, from the dining rooms to the hallways to the grounds, were institutional in scale and feel. Baby Boomers are going to be far more demanding of us as an industry, forcing us to re-imagine, re-design, and re-build from the foundation up. And this will include not just physical plants but the services, the experience, the *soul* of senior living. This upheaval in the industry will create even greater opportunities for those pursuing careers in it.

Another enormous challenge and opportunity for aspiring leaders will be addressing the void of options for America's aging middle-class. Financial models for serving less affluent seniors, except those at the bottom served by subsidized nonprofits, are virtually non-existent, though approximately two-thirds of the Boomer generation lack the pensions or savings to live in *today's* mid-market senior housing. Yet no matter whether one is rich or poor, urban-dwelling or rural, every older adult has fundamental needs that must be addressed, and they typically are highly influenced by how you live. Nutrition, physical fitness, mental fitness, socialization and disease management are the five pillars of well-being for older adults, and the industry will face the ongoing challenge and opportunity of addressing and providing them for its residents. Again, this will mean a greater emphasis on, and demand for, *innovative* leadership and management.

However, for us in the industry, it is easy to fall into the trap of inertia, because we are focused on executing our immediate strategies and delivering services to our current customers. But considering that we are servicing such a small percentage—just 15–20% of age-eligible and financially capable seniors—means we actually are vulnerable to those who envision a broader reach for the industry. What if a Google or an Amazon partners with some fellow technology companies to combine their products, their innovation, their skillsets, to create a business model that can be more widely applied? We know that tech companies and tech thinkers might be attracted to do so, because they like to engage in solving big problems, not just delivering stuff to your doorstep. Disruptive technology and disruptive management are both challenges and opportunities for the existing industry.

Disruption of a different kind and scale has hit our industry in 2020, and all its immediate and lasting effects cannot yet be fathomed.

Over my forty-seven years of leadership in commercial banking and senior housing, I have led teams through every significant challenge posed by a wide variety of economic cycles, both domestic and global. However, I have never encountered one as complex as what we faced with the COVID-19 pandemic.

Through the extreme market saturation years from 1999 through 2002, when "assisted living" was discovered by the capital markets, to the "Great Recession" years from 2008 through 2012, the tried and true solutions of raising liquidity, modeling optimal cash flows, and training innovative, solution-oriented leaders, were always highly effective in navigating illiquid markets. The signs were always there, for those who watched, providing lead times to plan and execute effective solutions.

The pandemic was a silent threat that came out of nowhere to collapse industries, rupture job markets, and throw the global financial markets into turmoil in the span of 60 days. The outcomes have yet to be written and there was no play book to follow. Solutions had to be crafted in days, not months, for something no one had ever encountered before. In the senior housing industry, we take care of the most at-risk population in the world, and we needed our teams to protect them while they themselves were exposed to great risk.

Our immediate focus over the first two weeks was to develop safety protocols that would protect the health of our residents and staff teams. A set of detailed plans were formulated to outline steps to be taken by our building teams. These were designed to guide the staff on how to create a safe environment for residents under different levels of exposure to the virus. These plans addressed all aspects of a resident's daily life at our communities, including dining, care services, activities, visitors, shopping, housekeeping, daily activities, and communication with family and friends.

The next point of focus was on our staff teams. We developed plans that outlined how to protect our team members and what precautions they should take for their own health, while caring for our residents. Those included daily symptom screening when they came to work, interviews regarding possible virus exposure they may have had outside the building, enhanced safety protocols inside the building, personal protective equipment requirements, tracking of contact points outside of our environment for anyone entering the building that would be interacting with staff or residents, and finally, additional compensation and benefits to recognize their efforts on behalf of our residents.

At a corporate level, resident and staff safety were given the highest priority; focus on profits was deferred. Communication was absolutely critical in the face of rapidly evolving awareness of how the virus spreads, as were best practices for managing resident and staff safety. With travel shut down, our team rapidly embraced daily Zoom and Microsoft Teams technology to connect community leadership, regional teams, and the corporate office. These daily video calls allowed our leadership teams to keep a handle on which teams were dealing confidently with the evolving protocols and which ones needed some extra support physically or psychologically to deal with the stress of the environment. The video call technology was extraordinarily helpful in raising the bar on professional performance and team confidence.

With safety prioritized, our teams began to focus on our resident's sense of well-being. Since regulations or internal protocols dictated isolating residents in their apartments, our operating teams began sharing creative ideas to virtually connect residents to lessen the effects of their sense of total isolation. We created games to connect residents in their apartments, coached communication around meal deliveries and daily delivery of care, and initiated fun models of enhanced socialization like our Rolling Happy Hour cart, which makes its daily rounds through the hallways connecting residents in adjacent apartments in safe, socially distanced ways. Above all we facilitated connections between residents and their families. We brought iPads and Google devices into our buildings and facilitated their use to visually and verbally connect our families who were remote or could not come into our buildings. We facilitated visiting days between families on a schedule that allowed us to keep a physical separation to ensure the safety of our residents.

As these safety and quality of life programs were adapted in the communities, we began a process of modeling the impact of mandated and voluntary restrictions on our business. As protocols evolved that shut our doors to visitors and new move-ins, the obvious impacts of revenue loss and expense increases were stress tested for cash flow and financial liquidity. Once we completed our projections for the immediate impact of the virus, we then moved to model the assumptions of a second wave of virus in the fall which might lead to another round of restrictions. This modeling was carried through to the potential cash flow and liquidity impacts on our business in 2021 and 2022. We assessed our ability to absorb the maximum financial impact and concluded that we needed to raise significant liquidity to support our business. To create liquidity, plans for discretionary capital spending and new investments were deferred or terminated. Those would be reconsidered once we had a clear picture on resources and duration of the pandemic. We deferred distributions to shareholders and cleared our bank lines to build both primary and secondary sources of liquidity ahead of any needs we projected through year end and into 2021.

At the same time our marketing teams were tasked with addressing messaging about the relative safety of communal living for seniors. There still exists today significant confusion about the risks of communal living in senior housing *versus* a nursing home environment. We have been capturing data daily on the incidence of infections within our communities, which has informed new social media and print programs about the safety of living in a Merrill Gardens community and well-run senior housing in general. We have collected print and video testimonials from families about their experience in living in a Merrill Gardens community for use in our marketing materials. All of these efforts have been focused on making sure our residents, their families, and prospects are well informed with an accurate picture of the level of safety they experience with us *versus* the risk of living at home and the challenges of trying to control exposure to the virus.

With safety, communication, lifestyle, and financial assessment programs in place, we have now shifted our focus to two very important elements of what we consider best practices. The first is to provide our teams with the steps and sequences to relax the safety protocols we have put in place while identifying which buildings will cross thresholds that allow them to reopen to family, visitors and new move-ins. The second step is also a best practice opportunity. We have been very successful in protecting our seniors and team members through the height of the pandemic, but what we put together to accomplish this were real-time crisis management solutions under aggressive time frames. At every Merrill building, we maintain a Crisis Manual that details the types of crises a team might encounter, from elopements to fires, floods, flu, and a series of other events, with quick reference guides on whom to contact, how to handle the safety of the residents, and even how to handle the media should they arrive on scene. But the one thing we had never envisioned was a crisis of this type, speed, and magnitude. While what we did worked, as we relax our protocols, we are also committed to assessing each step we took as a means of determining if what we crafted under pressure is truly what we consider to be the best practice. While the spread is moderating in the majority of our locations around the country, if and when it spikes in the months ahead, we will be prepared with a refined plan that has been thoughtfully assessed to be followed in the future.

While this has been challenging in every aspect of leadership, and while the ultimate path back to normalcy may well lie in the vaccines, we realize as a team that plenty of challenges remain before we reach that goal. I am extraordinarily proud of what our teams have accomplished, the solutions they have developed, and how creative they have become in protecting and keeping our residents active in the face of the pandemic.

Of course, even though dealing with the pandemic has become my highest priority, I must continue to practice the daily blocking and tackling of leadership. So, I have to continue to be mindful of what I previously had been seeking to improve in my leadership. Perhaps what leaders, including myself, must acknowledge above all is that we do not listen carefully enough to the voices around us, and that includes those of residents, families, and team-members, as well as our business partners and competitors. We can be assured that we will get off track or off mission when our communication is more outgoing than incoming, when we are talking rather than listening. In addition, too often leaders manage by looking backwards not forwards. As they say in the military, too often generals fight the last war. We get myopic about how things *have* been done rather than how they *should* be done. We get trapped in what our clients wanted as opposed to being engaged and understanding how their wants and needs are changing. While it is not unique to young leaders, it is more common among them to be so anxious that they let others know their point of view and do not broadly solicit the views and ideas of others. Solutions oftentimes are better arrived at by listening than talking or doing.

Complacency is the greatest sap on energy and effectiveness, and that applies from the individual to the institutional level. Personally, I need to constantly guard against defaulting exclusively to my experience—what is known in management as the Expertise Trap—in the decision-making process rather than considering relevant data, factoring in societal changes, listening to fresh ideas, and tapping into that competitive spirit which urges us to be better and achieve more. At 71, I do not have any thoughts of retirement, and a big part of what energizes me is the ability and ambition of the many bright young people working around me. To help them, and any others whom I can, become better leaders, is what drives me to continue to grow and stay young.

END

William (Bill) Pettit, President, R.D. Merrill Company (1993—Present)

Bill Pettit is the president and chief operating officer of the R. D. Merrill Company with responsibility for Merrill Gardens and sister company Pillar Properties. Merrill Gardens is one of the most respected assisted living operators in the country, with 70 communities in 20 states. Pillar Properties is an award-winning developer, owner, and operator of multi-family housing with 1,700 units in operation and the developer for all Merrill Gardens new communities.

Pettit joined the R.D. Merrill Company in 1992 after 18 years in the banking industry. During his banking career with two major regional banks, Pettit served in various roles including head of strategic planning, head of real estate lending, chief financial officer and president and chief operating officer.

Upon joining the RD Merrill Company, he was instrumental in the formation of the Merrill Gardens and Pillar Properties companies. Pettit directed the rapid growth and timely execution of acquisitions and developed the policies that speak to the Merrill Gardens and Pillar Properties commitment to quality. Pettit was a 2018 Seattle Business Magazine Executive Excellence Award winner and under his leadership the R.D. Merrill Company was named the 2020 Family Business of the Year by the Puget Sound Business Journal for its commitment to residents, team members and community service.

Pettit earned a Bachelor of Arts degree from Princeton University in 1971 and an M.B.A. from the University of Oregon in 1973. He was the first Senior Living Executive in Residence for Washington State University. He serves on the Argentum Board of Directors and he is the past Chairman of the Executive Board of the American Seniors Housing Association (ASHA).

Chapter 13
Changing the World: From Public Service to Assisted Living

Brenda J. Bacon

When I began drafting this essay on my path to leadership in the senior living industry, the world looked quite different from how it does today. It was 2019, and we had not heard of the COVID-19 pandemic, nor of the televised murder of George Floyd at the hands of the Minneapolis police. Therefore, as I write now in mid-2020, I must ask: What in the world is going on? What can we do about it?

I grew up in the 1960s, so naturally my plan was (and still is) to change the world. But sometimes, like now, that notion is especially overwhelming. How do you stop an invisible virus that targets seniors? How do you keep young Black men from getting murdered by "law enforcement"?

So, I could spend these precious few pages talking about my career, but a reader who is curious can Google me for those details. This essay—this book—is about *leadership,* and to be a leader you need to understand what problem you are trying to solve and be relentless in the pursuit of that solution. Plus, you must get other people to join you on that journey—people with the passion and sense of purpose that drive a successful team.

That is true whether the solution is aimed at creating the next, most wonderful version of the senior living industry; or keeping our senior living customers and team members as safe as we can from this vicious coronavirus; or helping our country answer the call to be a better nation when it comes to violence against young Black men.

I have always been drawn to the need for cultural change, whether it was access to quality healthcare, a quality education, the pursuit of the dignity of a job, or the basic values of treating people with respect, no matter their color, gender, sexual orientation, or age. After getting my undergraduate degree in Social Work, I went to work trying to eradicate generational poverty, one person at a time. You had to be

B. J. Bacon (✉)
Brandywine Living, Mount Laurel, NJ, USA
e-mail: bbacon@brandycare.com

M. Lifschultz, E. L. Schneider (eds.), *Paths to Leadership in the Senior Living Industry*, International Perspectives on Aging 29,
https://doi.org/10.1007/978-3-030-53966-5_13

there in the '60s and '70s to understand that what sounds like hubris today was what we believed we *could* accomplish and *needed* to accomplish then. Doing it one person at a time was too slow for me, however, and I turned to community organizing, where I worked with social service and healthcare agencies to improve people's lives on a larger scale.

This was a time of tremendous learning. Dealing with a variety of institutions, such as nonprofit agencies, private companies, and political bodies, I experienced how power and influence are applied to policy-making. I also learned the simple yet profound lesson that access to good healthcare and education impacts individuals, families, and entire communities. After getting an MBA at the University of Pennsylvania's Wharton School, I worked with a large hospital, then joined the senior staff of New Jersey governor Jim Florio and helped his administration reform healthcare, welfare, gun control, and environmental policies. Ha! As it turns out, nothing much had changed in terms of the size and scope of what I wanted to do. Florio was an inspiring leader and his courage has had a lasting impact on me. He always said, "We will do the right thing, not the easy thing." These words resonate with me to this day.

Working to reform welfare was perhaps the greatest challenge of my professional life; the most difficult and the most important. New Jersey was the first state to attempt such reform. Institutional opposition was fierce and came from actors ranging from powerful politicians to the faith community. But we had the conviction that allocating resources, such as education, training, and childcare to enable recipients to work, rather than continuing the existing, paternalistic system, was *the* way to end the cycle of welfare dependency. We succeeded, and our efforts became a model for similar reform throughout the nation. If you believe in a principle, then you accept whatever challenges you face to defend or advance it; you do not get talked into retreat.

Before entering public service, I had founded a healthcare consulting company and, along with some partners, designed, built and operated two nursing homes. Then, 24 years ago, I left public service, and, working with others, raised $65 million of private equity and started Brandywine. We built nursing facilities as well as assisted living communities, which were a relatively new concept at that time. Eventually, we sold off the nursing homes. Today, we have 29 assisted living communities, with about 3000 residents, in six states. I'm Brandywine's CEO and I *love* what I do.

Like public service, the senior living business presents the challenge and the opportunity to change people's lives. You have a solemn obligation to provide a safe, happy, and robust place to *live*, and not one where the residents sit around waiting to die. Yet there *is* resistance to the model, in that an older adult seldom raises their hand and says, "I want to go to assisted living!" Our residents are typically in their 80's, and we have dozens over 100. Many of them, and their families, might prefer to stay in their homes with a part-time caregiver. But that is not the healthiest choice, as we know that loneliness can be the greatest ager. We must create environments that welcome people and make them comfortable with putting their lives in our hands. It is gratifying on a daily basis to see that our residents and their families are glad they chose Brandywine.

We face, now, an additional challenge. Can our residents be social *and* safe in a congregate living environment in the age of COVID-19? The world will never be the same. A leader needs to recognize that and guide their team to drive toward that future and not run from it.

Fundamentally, I believe that to be an effective leader, you must have *followship*, which is defined as a team's willingness to accept direction and guidance from the leader or leaders of an organization. Merely having the title "leader" or CEO does not alone actually make you one. People must believe that you are worthy of following. Above all, you must demonstrate integrity and commitment of purpose. This cannot be overstated in our industry, where you are effectively asking your team to devote their lives to the care of other people, and often in stressful conditions for less-than-get-rich-quick pay.

I lead our organization with a deep and abiding commitment to our purpose, which can be understood through our stated core values:

- We believe that every resident has the right to be treated with dignity and respect and has the right of self-determination.
- We believe in delivering the best quality service possible to those who have put their trust in us.
- We believe that honesty, integrity, caring, and thoughtfulness are as important as the healthcare we provide.
- We believe that our residents are active, joyful, and fascinating individuals who will continue to learn and grow each day and will contribute to our learning and growing as well.
- We believe that great care is a team effort where ideas, collaboration, and mutual respect for each team member's efforts and contributions are essential to the result.

Leadership, it must be emphasized, is not a solo act. Our organization has many leaders and certainly among the very most important decisions I make is choosing who those other leaders are. Many of us in this industry believe that the executive directors of the individual communities are *the* key leaders, because people work for people, not companies. To have a successful community, the person directly running it must have extraordinary abilities to inspire, to manage, to problem-solve, and to create a vision and a sense of purpose for their team. Finding those people, grooming and training them, and of course, retaining them, are all crucial factors.

Indeed, we take leadership training seriously. At our Brandywine Center of Excellence, we train team members in the fundamental tenets of leadership, such as decision-making and teambuilding. We also train our people in cutting-edge clinical leadership so that, for example, we know our dementia residents are getting the best care in terms of daily engagement and motivation.

Leadership in our industry is becoming increasingly important, as the huge population of Baby Boomers reaches the age when our products and services will be relevant to them. Boomers will not be consumers of senior living as it exists today.

They will want a more customized experience that suits their lifestyle, which is relatively more active than today's older adults. We expect them to be more engaged in the communities where they live, beyond the four walls of their assisted living community, so that could mean volunteering, attending school, and even working. They will want more and varied choices, from the design and function of their homes to the overall food experience to the activities programming. The Baby Boomers have substantially changed every part of life they have touched up until now, and we should expect them to change senior living, as well. As an industry, we must be prepared to give them opportunities to live as *they* want, not as *we* think they need or ought, to live. Adapting to the new needs of an entire generation of customers will require deft leadership, at the individual level, the institutional level, and the industry level.

Leadership and management roles are of course already changing. That may be seen most prominently in the role of women, which is changing rapidly and dramatically, and which we actively support. Women at Brandywine are strongly encouraged—no, actually *required*—to speak up. No one has to defer to colleagues who are older, or have more experience, or even have seniority at the company. Young women, and perhaps especially young women of color, need to know that they are seen as being as smart and able and relevant as everybody else in the room. Unfortunately, tradition and history has taught some women to doubt their place as leaders, although that thinking will soon be utterly obsolete.

Another change happening is the role that Millennials are playing in the industry. As a cohort, they tend to have opinions on everything and are not shy in expressing them. Considering how large a percentage of the workforce Millennials are and will be, that *must* affect leadership and management. Principally, it means that as a CEO you had better be willing to accommodate dissent, debate, and democracy or you may have trouble attracting or retaining talent.

I take pride in Brandywine's forward-thinking organizational culture, which I believe is already positioned to take advantage of these changes I have mentioned. Historically, we have always emphasized respectful, challenging, and fun discussion around ideas for change. From this practice, we get pretty amazing results that energize the whole team. An added advantage is, when you get to the point of articulating and implementing a change, you will have the beauty of buy-in, not eye-rolling.

In addition, our organization is deeply collaborative. We are also very competitive—not with one another, but on behalf of Brandywine. We like to hear *everybody's* ideas. We debate everything. If we're in a room talking about matters ranging from building design to increasing market share, we have folks from all our disciplines in the room. We encourage a participatory work culture. We do not accommodate those who just idle along. We vigorously pursue change for the sake of improvement; who wants to do the same old thing all the time, anyway?

Our team lives by the doctrine that if one of us fails, we all fail. This all produces an invigorating environment, where creativity is welcomed, prized, and pretty much guaranteed.

So, our culture relies heavily on teamwork. And the chief rule for building and retaining great teams is, hire happy people. This may not sound very *original*, but it certainly is critical to your mission. Fundamentally, you cannot have happy residents if you do not have happy employees, and if your residents are not happy, your organization will pretty quickly decay. The best marketing, the strongest mission statements, even the nicest facilities, will not compete with the great results of high-quality work being done by happy housekeepers, nurses, program directors. Plus, hiring happy yields a virtuous cycle; happy employees help make residents happy, and they also attract other happy employees. As a leader, one should recognize that it is a lot easier to hire people who are happy than to train unhappy people to *act* as if they are.

Aspiring leaders are well-advised to take deliberate steps toward developing a knowledge and personal philosophy of the discipline, the *practice*, of leadership. I live by the belief that each of us meets someone *every day* from whom we can learn *something* if we have the humility to receive the lesson. A leader is defined by how they make people feel. As a leader, I consciously try to stay open to learning from everyone, whether a nurse on our team, a resident in our community, or a fellow CEO.

Of course, there are also formal ways of learning about leadership, such as reading. My favorite leadership book is *Leadership Secrets of Attila the Hun*, which sees leadership as the *privilege* of having the *responsibility* for directing the actions of others in carrying out the purposes of an organization, with *accountability* for both successful and failed endeavors. In senior living, you must take responsibility toward all of your constituents: your client residents, your employees, your partners, your investors, and the communities you operate in. You have the solemn responsibility to be a good fiduciary, to provide quality, and to constantly review your practices in light of current information.

Now, when it comes to forging that career path, people may be surprised by my strongest recommendation: first, "work the grill." That can be applied literally or metaphorically. Start out by getting a firsthand feel for the relationships that develop between the people who provide senior living services and the people who receive them. Those family-like bonds between employees and customers are unique to the senior living experience. Even healthcare settings, such as in skilled nursing or a hospital, do not quite compare, because those clinical environments are often transactional around care needs with little focus on relationships. Relationships are so critical to the culture of our communities and to the quality of life of our residents, their families, *and* our team members. To go directly from school to the corporate office, without getting frontline experience with residents, means that you miss getting a taste for this critical ingredient, which is a special part of the recipe for success in senior living.

While I have always believed in the value of such experience, this period of hand-to-hand combat with COVID-19 has just reinforced my belief. The disruption to our industry, which has come at the speed of lightning, means more than ever that leaders must know how things work from the ground up, not just in the C-suite. At the end of the day, nothing I say compares to the absolutely unbelievable performance of our team members on the front line taking care of our residents. Residents have not been able to see their own families due the isolation restrictions, but they could count on seeing that caregiver who may have come from a very different life experience than they have, who may, for example, be a single Mom who has never traveled the world but will fight every battle to get to you that day because you need her.

Indeed, it is imperative for leaders in our industry to appreciate staff at all levels. I often say to myself and to others, don't forget the people who give you the privilege of being a leader.

A privilege and joy for me is being around young leaders. Tapping into the wisdom of youth is like discovering a renewable source of energy. Hearing new approaches to problem-solving, ideas for new products, and new issues that we should be thinking about; all of this helps keep me sharp and informed and in touch as a leader. That is perhaps my chief strategy for personal growth as a leader. And I believe that this youthful source of energy is available to you as a leader no matter where you are in your career. That is especially true in these challenging times, when businesses need to approach the future with the energy and tenacity of a start-up. Going backwards is not an option.

As an African American woman executive, I am often asked, "What is that like?" Perhaps those posing the question are expecting to hear tales of woe that end in triumph. Or some secrets that I keep on how to navigate our challenging society. Sometimes, people do not find my answer to be very satisfying. And my answer is, I do not see race and gender when I see people; I just see people. And I operate on the mirror principle, that people do not see me as an African American woman, but rather as a serious and accomplished executive. The benefit of that is, being in a room full of white men, as I have been for most of my career, does not intimidate me. When you believe in your own abilities, you can be successful, whatever your individual gender, race, or age. Yes, you may have to work harder. Yes, you may have to experience an unenlightened person here or there. But follow your own passion and make it happen. You define your life. Don't let those who have lived by notions of what can't be done, or what is just too hard to do, to rent any space in your head.

END

Brenda J. Bacon is the President and Chief Executive Officer of Brandywine Living. Ms. Bacon co-founded the company in 1996.

Ms. Bacon served as Chief of Management and Planning, a cabinet- level position, under New Jersey Governor James J. Florio from 1989 to 1993. During President Clinton's first term, Ms. Bacon was on loan to the Presidential Transition Team as co-chair for the transition of the Department of Health and Human Services.

Ms. Bacon is a past board member and chairman of the Board of Directors of Argentum (formerly the Assisted Living Federation of America) where she advocated on behalf of the senior living business and the families they serve. She is the current board chair of the Senior Living Credentialing Commission.

Ms. Bacon is an independent director of two NYSE Boards: FTI Consulting, an independent global business advisory firm dedicated to helping organizations manage change, mitigate risk and resolve disputes: financial, legal, operational, political & regulatory, reputational and transactional, (NYSE: FCN) where she serves as chair of the Nominating and Governance Committee and serves on the Compensation Committee. Ms. Bacon is also a member of the Hilton Grand Vacations Board (NYSE: HGV), where she serves on both the Audit and Governance Committees. In 2013, New Jersey Governor Chris Christie appointed Ms. Bacon to the Rowan University Board of Trustees.

In 2017 Ms. Bacon was honored with the Virtua Health Humanitarian Award of the year.

Ms. Bacon received her undergraduate degree from Hampton University and her Master of Business Administration degree from The Wharton School of the University of Pennsylvania.

Chapter 14
Passion, Purpose and Perseverance: My Watchwords for Leadership in Senior Living

Lynne Katzmann

I was fortunate to have been born into a line of leaders, strivers, and entrepreneurs. My maternal grandmother was and remains a role model for me. In her native Germany, a woman had to have a dowry to marry, but her family was too poor to provide one. With her sister, she built a business, and my grandmother not only made a living, she also made herself "eligible." To honor her bold spirit, we tell my grandmother's story to describe our culture in Juniper's marketing materials.

Even closer to home, my father was an immigrant to America who rose to executive positions at several early technology companies, including Fairchild. Growing up in this milieu instilled in me a sense of possibility, not limitation. It gave me the confidence to believe that a woman, a first-generation American woman, could thrive in business.

My grandmother and father were also survivors. Perhaps as a result of their courage and determination, I have never viewed failure as an option. Instead, I have always seen obstacles as opportunities for honing my problem-solving skills, enabling me to convert what would otherwise be a failure into a success, or at least a learning experience.

Recently, there has been a fair amount written as Holocaust survivors battle COVID-19. Like my mother, they are in their late 80s and 90s. For her, the shelter-at-home requirements have been just one more adjustment to make in a life full of such adjustments. Gary J. Kennedy, Director of Geriatric Psychiatry at Montefiore Medical Center, who studies the mental health of this special population, says that resilience is common among Holocaust survivors. "(Resilience) is the norm, not the exception. Those that survive are a select group." As a child of survivors, I hope and believe that I inherited this attribute and that it positively impacts my leadership, particularly in this time marked by uncertainty, danger, and change.

L. Katzmann (✉)
Juniper Communiities, Bloomfield, NJ, USA
e-mail: lynne.katzmann@junipercommunities.com

M. Lifschultz, E. L. Schneider (eds.), *Paths to Leadership in the Senior Living Industry*, International Perspectives on Aging 29,
https://doi.org/10.1007/978-3-030-53966-5_14

We are all shaped by the world in which we live. I came of age at a time when women were taking on a new sense of self, and that influenced me, too; from youth into adulthood, I have always believed that I could do or be anything.

After college, I went off to Europe by myself and earned a PhD at the London School of Economics. Owing to my German roots, I studied the nation's 19th century healthcare system, and lived for a time in Berlin when that was a rather precarious situation. It was at the height of the Cold War, and the city was contained within Communist East Germany and divided by the Wall. Then when I returned to the U.S., I did not settle into the comfort of my "home turf" on the East Coast. I took a job in Oregon, working to develop their state-run health program. You could fairly say, my path was that of a bold and independent young woman.

Along the way, I developed a self-awareness that has served me well. For example, I am physically small, and I compensated for this by being assertive, which is a good trait. However, I could also be brusque, which might have been magnified because people often did not—and even today sometimes *do* not—expect this from a woman. I came to realize that it would be advantageous to soften that quality while retaining my assertiveness. In addition, I also recognized that my negotiating style could be overwhelming: I often achieved my objectives while convincing other parties to agree against their own interests. Such one-sided "wins" tend to leave trails of bad feelings, so, inspired by the classic book *Getting to Yes*, I learned to consider closely the needs of my counterparties and find mutually acceptable agreements. That is my approach to negotiating to this day, and I highly recommend the book.

I became a "self-appointed" leader at a relatively young age, as I founded Juniper in my late twenties. With big plans for myself and the company, but no track record as an owner or operator of senior living facilities, I set out to raise capital based on the principle that we would do well *and* do good by improving the lives of older adults. I emphasized the social importance of my mission, while also promising high returns for my investors. Not everyone bought my story, but I managed to find enough believers and attract sufficient financial support to launch the business and, crucially, to inspire confidence in my leadership such that I retained voting control of the corporation, which I have to this day. As the founding CEO, I retained total flexibility and authority to make decisions, but also sole responsibility for the results. So, I assumed all the burdens of leadership without much formal experience or training. And I have been learning ever since.

Fortunately, I have had resources to draw on in the form of *people* I have known, *principles* I have studied, and a *philosophy* I have adopted.

In terms of *people*, several were key to my development as a leader. Above all was my aforementioned grandmother, whose story as a trailblazing businesswoman was, and remains, an inspiration to me. As a child, I really admired my best friend's mother, who was a full-time practicing physician *and* a full-time mother with four children; of course, that is a lot of responsibility, yet she got it all done. I still think of her strength and style. Then, early in my career, I worked for a great executive and social entrepreneur, Thomas J. Higgins, from whom I learned so much about a leader's foremost responsibility: good decision-making. Finally, I learned the importance and intricacies of capital structure from John Kidde, a brilliant financier who served on our board. Because of the profound influence of these people in my life, I would urge all aspiring leaders to actively seek out mentors and role models from different places and disciplines; I did, and it has made a significant impact on me and my career.

I learned profoundly important *principles*, and validated ones that I had already been applying through my leadership, from the classic book, *Good to Great* by Jim Collins. Since being published twenty years ago, this book has been a veritable business bible for leaders aspiring to create and run great companies. My advice: read it.

Two of the principles Collins identified especially resonated with me: Level 5 Leadership and The Hedgehog Principle. Level 5 leaders, Collins says, "display a powerful mixture of humility and indomitable will. They're incredibly ambitious, but their ambition is first and foremost for the cause, for the organization and its purpose, not themselves... (though) Level 5 leadership is not just about humility and modesty. It is equally about ferocious resolve, an almost stoic determination to do whatever needs to be done to make the company great." When I read those words for the first time, I saw myself described, and still do when I re-read the book now.

The Hedgehog Concept was equally meaningful to me. With this, Collins describes how leaders of great companies focus passionately on "not a goal to be the best, a strategy to be the best, an intention to be the best, (or) a plan to be the best, (but an) *understanding* of what you *can* be the best at (emphasis supplied)." Further, he elaborated, if a leader then makes a series of consistently good decisions that lead to "supremely well executed" actions "accumulating one upon another, over a long period of time," a great company can be built. It is with such singular focus and commitment to be the best that I have always run Juniper, specifically in delivering the best possible service to our customer residents while earning the best possible returns for our investors.

Philosophically, I believe in practicing servant leadership. Popularized by Robert Greenleaf's book of the same name, to me servant leadership means doing whatever I can to help everyone on my team—from our line workers to our executives—be the best they can be at what they do. It means doing whatever I must to optimize the organization for the good of our residents, their families, the community, our investors, and for society as a whole. It means considering these priorities with every action we take. That might mean, for example, visiting one of our buildings and doing a shift in food service, either because they are short-staffed or because that is just a great way to get to know our people. Or it might be innovating an integrated care program like our "Connect4Life" that will be good for our residents and my company, then sharing it with industry colleagues and competitors for the benefit of potentially millions of older adults.

Juniper's non-hierarchical, or flat, corporate structure also is an outgrowth of my leadership philosophy, part of which is to value humility, in myself and in our people. When people are passionate about their work, feel suitably rewarded for it, and put purpose over ego, vanities like job titles and where people sit on an org chart cease to be anything but *obstacles* to success; they certainly do not serve the interests of success. Many of my team members have been with Juniper for ten to twenty years, and it is a testament to their sense of purpose, both individually and collectively, that we can have this kind of corporate structure *and* have high rates of employee satisfaction and retention.

Indeed, it is critical for leaders to be humble, especially in our industry, where your goals truly are—or should be—all about serving others. Humility does not mean ignoring indicators of success, however; indeed, it requires an honest and constant accounting of those indicators. I closely watch three criteria, which give me the feedback I need to know if my leadership is effective: finances, culture, and innovation.

With regard to finances, it is hard to argue with results. Ultimately, either you can meet payroll and grow your business, or you cannot. You can satisfy your customers, or you cannot. You can provide great returns to your investors, or you cannot. I am proud that Juniper has met and exceeded our standards in these areas since our founding, holding ourselves to a "double bottom line" of strong financial results for our investors and superior quality of life for our residents. This illustrates the concept of doing well by doing good.

As to culture, the way we apply an indicator is to ask, do our mission and values truly translate into practices? At Juniper, they do. Our mission is to nurture the spirit of life, and in everything from the design of our buildings to our enrichment offerings to our programs of care, we have always been true to it. It is a quality that can and will be readily experienced by visiting any of our communities and seeing how team members work and observing how our residents *live*. Our vision, since our founding, has been to change the face of aging, and that comes to life in the excellence of our innovation, which of course is the basis for positive change.

Juniper has a long track record of innovation, and our most recent effort may be the most promising and impactful. Our program Connect4Life integrates clinical providers, their services, and communications protocols to provide individualized, coordinated care to our residents. We consider it a marriage of on-site clinical services and the supportive environment of our communities. It is meant to reform the highly fragmented, duplicative, and non-economic care system that too many older adults experience, one that often leaves them out of the decision-making and results in low satisfaction and poor outcomes. Connect4Life is a high tech, high touch lifestyle management program that we believe will significantly improve the lives of those who participate in it. And we have estimated that by sharing it with our friends and competitors in the senior living industry, Medicare could save $10 billion annually. So, I am proud that Connect4Life may well become not just a beneficial program for our residents but also a widely adopted legacy of Juniper's commitment to innovation.

The clarity of our vision, which is best demonstrated through our innovation, also helps me minimize making perhaps the most common mistake leaders make: a failure to adequately communicate a company's "why." Of course, what that means is, its *purpose*. As opposed to talking about *what* you do and *how* you do it, which tend to be forgettable and uninspiring—like the text of an instruction manual—when you share your purpose, people will not only pay attention, they will remember what you say. Too often, leaders get bogged down in the crisis at hand, or ongoing operations, or financial projections. Such "how" and "what" topics do not motivate your team or create strong bonds between you and your stakeholders. When you discuss your purpose, however, you are sharing your values and passions, and humans are hard-wired to support such aspirational thinking in others. Because our vision is so much a part of who we are at Juniper, I devote a good deal of my time to communicating our why, thereby fulfilling one of my chief responsibilities as a leader.

If I were to identify the other most important role of my leadership, it would be taking care of what I would affectionately call my Juniper friends and family—our stakeholders. And though I know them very well, I can always know them better. When I feel the need for a "shot" of improvement in this area, I always know where to go and what to do: I seek out our stakeholders and I *listen*. That may consist of one person, or a small team, or a large group; it could be in a private setting such as coffee with a board member, or a public setting, like an all-hands meeting of team members at a community. Of course, residents and their families are always those that I listen to most carefully; without them, there is no company. I believe this takes adjusting for some leaders, as popular culture conditions us to think that it is only the leader who does the talking. But, as the Dalai Lama said, "When you talk, you are only repeating what you already know. But if you listen, you may learn something new." Whenever I go out listening, I always come back with new information, new ideas, new inspiration. And I believe that listening as a strategy for developing leadership skills is equally valuable to young, would-be leaders.

Along with listening, I believe there is no substitute for spending time in the field, whether you are an established executive or an aspiring one. One simply cannot learn everything there is to know—especially in this highly *personal* people-driven industry—from studying books or sitting in an office. One must spend time "in the buildings" to get as close to the experience of your customers as you can. Reading a blueprint may tell you how wide a hallway is, but seeing a resident struggle to get around that table in the foyer will tell you how wide it *ought* to be. As I remind myself and gently admonish others, "You won't get it until you get there."

However, while there is always a rich vein of information to be mined from within your own company and its communities, leaders must also learn from outside the industry, and do so actively, and on an ongoing basis. It cannot be overemphasized that in this industry we must recognize how big a role we have in people's *living*, and that what is going on outside our four walls must be considered and integrated into our products and services. Perhaps most salient now is the whole gamut of available technology—whether and how we can or even *must* put it to work for the sake of our residents (and employees and, yes, our investors). For example, I keep a close eye on Amazon, which plays an increasing role in so many of our lives through its Prime services and its voice-activated smart home features. It is conceivable that the one-time bookseller may end up a major player in senior living. As leaders we must keep a close eye on what is happening around us.

I believe this has implications for how leadership will change in the senior living industry. While the industry has always presented challenges due to being highly regulated and people-focused, there also has been a certain amount of predictability to it. The business can be driven by process and tactics, like pricing and occupancy. Indeed, the crude expression some have used to describe the senior living industry is, "heads in beds." The plus to this has been that the product and service offering has remained relatively stable. But with the generational shift that the Baby Boomers will bring, the industry will soon be driven by *strategy*. Leaders will have to devise strategies to meet the needs and demands of a population that will defy monolithic categorization. Boomers will live longer, more engaged lives, and the onus will be on those of us in the industry to figure out how to accommodate that and do so with perfect timing. Meanwhile, the use of technology—most importantly, big data—will no doubt accelerate, and offer us opportunities to improve care, enhance living arrangements, and widen programming options. But all of this will happen rapidly *and* while we are still running our existing businesses for our paying customer residents. This is going to require leadership that is open to change, able to innovate without fear, and to operate nimbly. Oh, and did I mention that demographics and economics will drive changes in revenue structures, as well as in locating and building physical spaces? Leadership in senior living will not be getting any easier.

Indeed, 2020 brought the greatest challenge to leadership in the history of our industry: the tragic COVID-19 pandemic. The virus has shaken the foundations of our society, our economy, and our health system, and has found its primary victim in the older population from the moment it arrived in the United States. This fundamentally challenged how I practiced leadership.

Juniper has been a flat organization that honored collaborative decision making. Because of the uncertainty and the speed of the virus's transmission in hot spots in which several of our communities are located, this situation called for a different style of leadership. As the leader, I found myself needing to listen but then act decisively—sometimes without the full consensus of my team. I would characterize this as strong leadership, which I define as clear, decisive, and proactive.

This strong, strategic leadership approach is exemplified in Juniper's COVID-19 testing strategy. We made the decision on March 27, 2020 to test *all* residents and team members. We designated an individual to focus solely on testing. She identified an FDA-approved lab that was willing and able to supply us with at least 3000 tests and then process them quickly and provide HIPPA-secure results. The labs provided a 24–48 hour response from the time the test was received at the lab. This meant at least 72 hours for turnaround, but in actuality, it took up to 7 days in rare instances. In order to test, we needed a prescription and a licensed professional to do the testing. We had both, which is not always the case in senior housing.

We then developed a policy and procedure for testing which detailed the way the test was to be administered, including where the testing would occur, PPE, storage and shipping, and other details. In addition, we drafted consent forms to be used for all individuals being tested. In short, we identified a source, designed a process, and trained staff within 5 days, which was exceedingly fast and illustrative of our team's effectiveness. We boldly began testing on April 1, at a time when the standard practice was to test only highly symptomatic people, primarily those in hospitals seeking care. We purposely used a private lab and of course paid for the testing of all associates.

It was our longstanding commitment to innovation that made us willing and able as an organization to search for new processes and protocols that would protect our residents and associates. We found that the willingness to be strategic even during times of great change provided staff and families with additional energy to "stay calm and carry on," as they felt that our work on testing was "leading the charge." This thought leadership, coupled with quick and detailed execution, gave us the ability to create our own fortune. We devised a plan with actionable steps that could be celebrated along the way; it gave people comfort that there is something within their control that will have positive impact.

The results largely contradicted what we had been told by certain authorities. Between 70% and 94% of all individuals who were tested, including both residents and associates, were asymptomatic at the time of testing, and the vast majority remained asymptomatic. Of those who did become ill, most recovered. By being proactive, we learned that the real danger was in asymptomatic and pre-symptomatic spread. We also learned that the arc of the disease was approximately two weeks.

In our higher acuity communities, where we had standalone memory care residents for example, we chose to protect the teams by creating a shelter-in-place "bubble" where teams worked alternating 12-hour shifts and lived on campus, thereby eliminating the possibility for external exposure and asymptomatic transmission. Where there was significant though largely asymptomatic virus activity, we created small geographic cohorts, or neighborhoods, with dedicated staff to limit the number of individuals who shared contact.

By late April 2020, our communities with cases were largely recovered, meaning that no residents or associates were active carriers according to CDC guidelines. Our early plan assumed that within a month, rapid testing would be available. It was not and so, again, at least one week before any similar program was announced within the industry, we began regular, weekly re-testing of associates. Our short-term goals remained the same: until a vaccine was available, we would test all visitors, including staff and residents who left the community and then returned. To that end, we actively worked with world-renowned MIT scientist Feng Zhang, who together with colleagues had developed an FDA-approved test using CRISPR technology, which not only speeds up antigen testing but also appears to have high levels of accuracy and will be far less expensive than existing polymerase chain reaction (PCR) tests. As of this writing, we are working via Sherlock BioScience (of which Feng Zhang is a founder) on a device to process tests in remote, non-laboratory locations. The latter would be perfect, though it is not immediately available. In the interim, we are hoping to use local or regional labs with existing approvals for processing complex tests to enable us to use this test regularly to significantly limit the chance for disease transmission inside our communities.

Being willing to lead, to analyze other areas and countries for their experience, and to apply those examples with clear and decisive action is what has characterized our response to COVID-19.

Of course, there are developments in the industry to look forward to as well. I must emphasize that at least one inevitable shift is right in front of all of us, and it will have a tremendous, positive impact both in our industry and in U.S. business generally: the increasing percentage of women in leadership positions. If we want to improve business in this country, all we must do is make sure that 50% of all corporate board members and CEO's are female. This notion may raise an eyebrow or two, but to me it is that simple. There is ample research that demonstrates the positive effects on businesses with female leadership. And that does not necessarily mean that every woman is or must be a "feminist." It is simply that women are good businesspeople, and often different from men. And perhaps it results from mothering, or the mothering instinct, which brings a different dynamic and sensibility to the workplace. At the board level, when women are involved in decision-making, we see shifts in priorities and goals as well as how they are pursued. As we see more and more women joining the executive ranks of senior living companies, I believe we will see a better industry emerge. I strongly encourage professional women to pursue careers in senior living, and I like to say to them, "There is plenty of room at the top for you!"

Aspiring leaders often ask me for advice. I share with them a few points that I believe have served me well and that have a universality to them, like the basic laws of math or physics:

1. Lead with your beliefs and values; find a place in this big world where you can do so. Trust me, there is a lot of room for you.
2. Like they say in politics, stay connected to your base; know the hopes, dreams, desires, and needs of your customers, your team, your community.
3. Remain open to learning and seek learning vigorously; learn about what *is* right in front of you, but with equal vigor learn about much that *is not*.

END

Lynne Katzmann, PhD founded Juniper Communities, which invests in, develops and operates senior living and long-term care communities at the age of thirty-two. Bringing with her no financial backing, a mere six year's business experience and a PhD in economics, her mission was to improve the last years of life for seniors.

Katzmann believed that a woman could and should lead this business where the customers—staff, residents and family caregivers—were predominantly women. Her goal was to make Juniper profitable for investors; and as a founder of the then nascent social corporate social entrepreneurship movement, she would adhere to the double bottom line: doing well by doing good.

Today, 32 years later, CEO Katzmann continues to actively lead Juniper, now with 22 properties in three states, more than 1550 employees, and $85 million in revenues. Juniper is ranked #12 by Crain's New York on its annual Top 50 ranking of woman-owned companies for the New York tri-state area.

Juniper's investors have enjoyed double digit annual returns while her residents, their families and Juniper Associates give the company high marks on annual customer satisfaction studies. Health-related metrics, such as low hospital readmissions, support Juniper achievements with hard data indicators.

Katzmann received her undergraduate degree from Tufts University and her PhD from the London School of Economics.

Juniper is based in Bloomfield, New Jersey.

Chapter 15
Leadership in Aging-in-Place: Apply the Golden Rule

Paul Kusserow

I grew up on a hill farm in Northern Vermont. We lived outside a small village under the State's tallest mountain, Mt. Mansfield. I spent all my time outdoors—skiing and playing pond hockey in winter; fishing, playing games in the fields and woods during the summer. It was a small town, natural world. It was a Norman Rockwell childhood.

The family business was academia. My mother was a nurse who, after my dad died, got her PhD and became a Fulbright Fellow. My brother and sister are both accomplished academics. My father was a physician and a professor at the University of Vermont. He left a position at Yale to raise his family, and do research on the artificial heart, in the idyllic Vermont countryside. Dad was killed in a car accident when I was 14. This crisis caused our family to lose its gravitational pull and we all drifted. I learned the hard way that life can change quickly and dramatically.

I managed to receive a scholarship to attend prep school and did well enough there to attend Wesleyan University. In college I was a seeker and a learner: I was involved in varsity athletics, editor of the school paper, and active in student government, along with studying religion and philosophy. I won a fellowship to study poets in New Zealand, and another to document the characters who inhabited a soup kitchen where I volunteered. I was selected to be a Rhodes Scholar and attended Oxford, where I read at least 500 books.

So, I was aiming for an academic career, but after experiencing three years of solitary study at Oxford, the idea of getting a PhD became unpalatable. While considering a career in journalism, I learned that McKinsey, the prestigious management consulting firm, was looking to diversify their personnel profile by recruiting non-MBAs and seeing how they performed with in-house training. When I got an offer, I accepted—mainly because they paid well, and I could learn something I

P. Kusserow (✉)
Amedisys, Baton Rouge, LA, USA
e-mail: kendra.kimmons@amedisys.com

M. Lifschultz, E. L. Schneider (eds.), *Paths to Leadership in the Senior Living Industry*, International Perspectives on Aging 29,
https://doi.org/10.1007/978-3-030-53966-5_15

knew nothing about. I was not a great McKinsey associate and I found the go-go world of 1980s corporate life distasteful, but I received a great business education, and I came to learn and emulate the way strategists analyzed and solved problems. From there I moved through a number of strategy and development jobs in publishing and the nonprofit sector, learning a lot but not achieving much success.

Eventually my very patient wife got tired of moving around and suggested we move to Santa Barbara, her favorite place. I could make something work there and we could raise our family there. Eventually, I was introduced to Jeff Barbakow and Trevor Fetter, then the CEO and CFO, respectively, of Tenet Healthcare, the owner/operator of hospitals, which was headquartered in Santa Barbara. They were great mentors and very supportive of my professional development, letting me create a venture fund to incubate and build companies. Through investing in innovative projects at the vanguard of healthcare, I found that my professional home was in this field.

After Tenet, and stints at the Advisory Board and Humana, I had established myself as a strategist who understood the healthcare ecosystem and could analyze situations and devise effective business solutions. As a result, I had been given roles in strategic planning, business development, mergers & acquisitions, and innovation. I had also run portfolio companies and business units, sat on numerous boards, and successfully completed a couple of small turnarounds. It was a solid career, to be sure, but I was in my early 50s and had never run a standalone company; being CEO had eluded me. The top ranks in healthcare were and are dominated by operators and financial people. The conventional wisdom was, and still is, unless you were an operator, or maybe a finance whiz, you could never run a company in the industry.

Just as I was helping to launch a start-up with a major private equity fund, an executive recruiter presented me with an opportunity to be a CEO. He was quite blunt in saying that, if I did not take this role now, I would soon age out and never get another chance. If I was going to run something, now's the time.

The company was Amedisys. The company's leadership and strategy were failing, and after a board-induced management shake-up, they tried to convince a number of industry leaders to run the company but did not succeed. By the time they got to me, the CEO position was radioactive; no one wanted to touch it. I was probably their last choice, but I was willing to take it on because I thought underneath the misguided exterior were great bones.

I also had a personal principle at stake, one which I was eager to prove in business. Before Tenet, and a couple of times since, I had worked for a lot of bad bosses and could see why so many big-time players in business had bad reputations. I had concluded that many at the top are actually not particularly good people; they are far more interested in accumulating wealth and personal glory than improving the lives of others. Too few believed that when delivering a service or a product, your job is to serve *that* cause, not your ego and personal desires. When *you* and *your needs* are the priorities, you are being contemptuous of your employees, your customers, and the cause you are meant to serve. When that happens, you will, inevitably, hasten your decline. That's why many companies do not last that long.

The principle I wanted to test was the Golden Rule: "treat others as you would want others to treat you." Found in virtually every religious and ethical tradition on earth, it is generally thought of as a guide for heightened human relationships. Ever the strategist, I wanted to apply the Golden Rule as an overarching corporate strategy. I wanted to show that if an organization takes great care of people, treats them as we ourselves would want to be treated, that company would thrive. *That* was my motive for taking the job.

When I was at Tenet, I got to know a wonderful man, Fred Reichheld, who had written a book called *The Loyalty Effect*. He helped me crystallize my thinking. His thesis was, if you treat others—employees, customers, stakeholders—the way you yourself want to be treated, if you are in their shoes, you will significantly differentiate yourself and business will boom. Perhaps because of my studies in theology, I agreed. I believed that if we served our patients and our employees well, we would become a phenomenal company. We would take great care of our patients and achieve excellent clinical outcomes. We would attract the best employees and minimize turnover. We would stimulate productivity, because caregivers want to be able to take care of people, and if you create the right environment for that, they will do so heroically, *every day*.

In 2014, our company faced one cascading crisis after another. We paid a $150 million settlement to the U.S. Department of Justice for improper Medicare billing. Our founder and CEO resigned. We closed care centers and laid off hundreds of employees. Our stock sank sharply, and staff morale plummeted to an all-time low. Since mid-December 2014, when I stepped in as CEO and started to change our culture and the way we take care of people, Amedisys has been reborn. To date, our company's market cap has grown seven times so far (May 2020). Our workforce has grown almost two times. Today, Amedisys is the country's second largest provider of home health care and third largest in hospice services, with care centers in 39 states and the District of Columbia. Last year we made more than 12.3 million visits to care for more than 415,000 patients.

In the 20th century, elderly adults facing health challenges generally opted to live in nursing homes, retirement communities, and, once the concept was developed in the 1980s, assisted living facilities. But in the course of a generation or two, a sea change has occurred, as older adults increasingly prefer to stay in their homes and receive healthcare services there when possible. Approximately 1 of every 25 Americans—an estimated 12 million—currently receive health care at home. This is known as "aging in place," and it can properly be described as a *movement*. According to an AARP survey, 88% of adults age 50 and older say they want to remain in their homes and communities as they age. The Centers for Disease Control and Prevention define it as, "The ability to live in one's own home and community safely, independently, and comfortably, regardless of age, income, or ability level." Given the demographics, Amedisys is well-positioned for the future. Americans are living longer than ever before, and a higher percentage of our population is over age

65. In 1960, only 6.4% of the U.S. population was 65 years of age or older. Today it has more than doubled, to 13%. This trend is only going to accelerate in the decades ahead. By 2050, for example, that figure is predicted to nearly double yet again.

Amedisys is in the business of providing home health, hospice, palliative care, and personal care services. Our average patient is about 77 years of age, suffers from several chronic illnesses, takes an average of 13 medications daily, and regularly sees multiple medical specialists. They are highly vulnerable, with health issues including cancer, diabetes, heart failure, high blood pressure, COPD, depression, and dementia. Some require management of chronic conditions, while others need post-acute care after surgery, rehabilitation from an injury or, in the case of terminal illness, palliative and end-of-life care.

It takes a special caregiver to do this intensive clinical job day in and day out. Our professionals do everything from taking blood pressure to managing wounds and administering physical, occupational, and speech therapy. Our teams work together to find options for the care that best fits each patient. They also make sure patients eat properly, dress and bathe themselves, move, ambulate, and take the right medications at the right time. Our teams arrive at the doorstep and give the most intimate one-on-one attention, including that most crucial of intangible elements, namely compassion. They smile, lend a sympathetic ear, and pat patients on the shoulder for encouragement. As a result of this expertise, many patients eventually move better, eat better, breathe better, speak better, suffer less pain, and even live longer. They gain the opportunity to preserve a sense of freedom, independence, and dignity. A woman once unable to walk again climbs to the top of the stairs. A man who lost the power of speech tells his wife he loves her. This is what we live for.

Home healthcare primarily aims to prevent patients from being either admitted to hospitals in the first place or readmitted, and also from having to enter nursing homes, assisted-living facilities, or rehab centers unnecessarily. Accordingly, healthcare delivered at home improves patient outcomes, cuts overall health care costs, lowers hospital admissions and readmissions, and accounts for numerous emotional and social benefits that can only be gotten at home.

In our view, the bottom line is this: healthcare should always be provided in the setting where it can best meet patients' needs and goals. Healthcare at home performs at its best if properly coordinated with all other clinical collaborators, whether primary-care physicians, hospitals, or other institutions. Our main mission at Amedisys, in short, is to better enable Americans to age in place, optimally in the place they want to be.

So, to what do I attribute the company's success? *Our strategy*. By collectively devising the right strategy, proving it's the right one, implementing it, nurturing it, articulating it, and constantly revising and perfecting it, Amedisys transformed itself.

The strategy we ultimately forged came straight from our people and our patients—the grassroots—a trickle-up approach, not top-down. When I first started here, I ventured into the field to watch and listen to thousands of our employees in action, as well as our patients and the providers that referred to us. I came face to face with our clinicians—nurses, therapists, social workers—and also our patients in their homes and at the bedside. I listened to providers who referred to us and those who did not. As I've noted, I had developed a lot of strategy over the years, but as a first-time CEO, and for the first time in my career, I decided to actually have our employees, patients, and referral sources, in effect, write our strategy, and they did it better than I could have done.

I approached it that way because, though I had often worked directly with CEOs, I had not appreciated just how different things are when *you* are the CEO. This is manifested mostly in the degradation of your ability to get good intelligence; that is, people simply stop telling you the truth. So, to get at the truth, I visited our care centers around the country and met with thousands of our employees, patients, and referral sources before they knew who I was.

In short, I made house calls on the very people who themselves make house calls. I wanted to understand who our staff are and what they do. I asked, "What's really important to you? What do we need to do to be better, different? How can we deliver great care?" And I asked our patients, too, as well as those who recommended or would not recommend us. Whenever I came back from those visits, I would always have a clearer vision of what we would need to do. It always validated my sense that our employees, our patients, and the providers who used our services would be the best source of our company strategy. My goal, then, was to hear from the people on the front lines what was working at our company and what should be improved. In other words, I led by listening and tuning our company to the needs of the market.

I gathered all the feedback and distilled it into an ultra-simple four-pillar strategy: 1. Become an employer of choice. 2. Achieve clinical distinction. 3. Pursue operational excellence and efficiency. 4. Do all three of these and you grow. Implicit in the strategy, of course, is, always practice the Golden Rule.

Being an employer of choice and having the best staff possible feed the other three strategic imperatives. Our priority is to enable employees to spend the most time serving patients at the points of care, and to do so consistently and effectively. The better our clinicians perform, the better our patients do, the more providers refer to us. Loyal employees and great patient outcomes help enable an excellent and efficient operation with strong productivity. All of these factors drive growth. My job as CEO, in essence, is to be a caregiver to our caregivers.

Our intent has been to promote a culture that engages our staff more deeply, retains these employees, grows our business, and unleashes our potential. A compelling narrative for your people is a *sine qua non*. People want their work lives to have meaning. If they believe in what they are doing and see their efforts as part of a worthy strategic vision, you will get greatness out of them. So, I spend much of my time telling our story, bringing people in, showing them how important their contribution is, and highlighting how their specific actions enhance our collective strategy as well as those actions that detract from it. We also solicit *their* view of the story and therefore can measure that in our overall engagement score—which reflects employee pride, advocacy, loyalty, and satisfaction.

We seek ways to demonstrate pride in our work. For example, we have an internal initiative called "Why We Do What We Do." Its purpose is to remind all of us—clinical and non-clinical staff alike—about the stark realities of caring for the sickest of the sick and, in some cases, the dying. To connect the dots between our mission and our strategy, we photographed images of our employees caring for patients at home and in hospice—photos that were distributed and displayed throughout our offices. Such an effort continually illustrates and reminds us of the importance of the work our people do.

One thing that's important for Amedisys is for us to know we are all caregivers. One way or another, whether you are in accounting, or you are a nurse, whether you are in payroll or you are a hospice chaplain, whether you are touching patients directly or not, you are contributing to our efforts to provide care for people. And that is one of the most important things anybody can do in their life. Connecting us all as caregivers has really unified us. Internalizing the commitment that we put care before anything, and we will never compromise the high quality of that care, has, I believe, deepened our sense of mission and purpose.

We also continue to invest heavily in professional development and orientation and training programs. We award cash bonuses for clinicians and support teams at the Amedisys home health and hospice care centers that receive the highest quality ratings. All of this has raised morale higher than ever before.

This strategy never rests; it is ongoing. We're still not there, and probably never will be. I still do field visits to stay directly in touch with clinical staff and patients alike. That practice has never stopped and never will. It is how I determine, more accurately than any other metric available, whether our strategy is working or is not. It also gives me the chance to say what I most need to say to our staff, namely, thank you for the work you do, and how can I help?

So, I spend about a week per month out in the field. There is a certain spirit that I try to cultivate among our staff through my time with them. It comes down to this: the answers are within us. The best is here, but we have to find it and then spread it around. This accomplishes two things: showing the organization we have excellence everywhere and giving our own people credit for it.

Currently, we are in the midst of the COVID-19 pandemic, and our industry, like so many others, is having to adapt and evolve daily to meet the challenge and continue to provide the very best care to our patients under very limiting circumstances. This is uncharted territory, and it takes the creativity and ingenuity of our full team to maintain our commitment to our mission while keeping our people and our patients and their families safe.

Personally, I have spent much of my time on the regulatory and legislative front, trying to make sure we have a favorable environment to operate in. Our team has been incredible in rising to this challenge, communicating effectively with our clinicians on how to continue providing care in this new environment and equipping them with the tools they need to do so.

I think we are going to come out of this crisis culturally and operationally stronger and with a deeper relationship with our employees. I also believe the industry will come out stronger if we are doing the right thing with our policies and procedures. Never waste a good crisis.

Though I am the titular leader of my organization, I am not remotely a one-man show. One-man shows are not possible in this environment. They are also ruinous. My view is that a relay team can get around a track faster than a solo runner. Yes, there are exceptions, like Bill Gates and Steve Jobs—those who seem to be able to put themselves in the center of an organization and do everything. And early on, when we were in turnaround mode, I had to make unilateral decisions. There are different leadership needs for different times. CEOs ruin good companies by not seeing that a company has evolved, and their particular skills are not applicable anymore. When I became CEO, the company's survival was at stake. I had to make lots of quick, solo decisions. In crises, often leaders must make decisions on their own.

But as we stabilized and cleared the choppy waters, I realized early on that the less credit I took, the fewer orders I gave, the less I insisted on elevating my opinion, and the more I opened up the discussion and debate at a table stocked with people smarter than I am, the better we did. Now, I no longer make singlehanded decisions and have not done so in the past last few years.

So, when it comes to leading, I believe it is optimal to create the conditions under which a whole emerges that is greater than the sum of its parts. That is done by teamwork. Thus, I keep a keen eye on building and maintaining my team. And I do have a superb team. In fact, I view them all as better performers than I am. We like to combine healthcare generalists with those who have a very specific domain of knowledge in our core businesses because we can generate creative ideas but keep them grounded in operational viability. This allows us to be operationally excellent in our care but have a continual push to innovate and change. My job is to bring the team together, put the big problems on the table, and collectively work out solutions. We then divide up the responsibilities, track the results, and constantly correct course until completion. The fact is, individuals cannot scale themselves enough in a company this size to produce consistent results; the whole company has to work as one to produce spectacular results.

Putting together a great team is a form of alchemy, and though there is a special quality to the magic that you cannot entirely control, you can strongly influence it with your intentions; that is, who you hire and why. I have a particular point of view on this. Until I became CEO, I felt that I had won many silver medals in my career, but never a gold. I experienced it as a vague and looming dissatisfaction, not getting quite the opportunity I wanted to build an organization that worked well with the principles I thought were most needed in an organization. If I'm anything, it's stubborn and resilient. I hung in there and got the opportunity to fix a broken company. I got my chance, and I'm trying to make the most of it.

As a leader, I have made it a practice to build a team of silver medalists, those who achieved well enough but were never wholly acknowledged for their talent nor had a chance to really show what they could do, almost always due to no fault of their own. Those who have missed in this way—to reach the top, to win the gold—approach a new opportunity very differently than those who have risen quickly and never failed. I believe that these silver medalists so appreciate getting the break they thought they'd never get, that they will not let you—or themselves—down. No matter what, they will not fail. They seize the opportunity and use it as motivation to show all the people who passed them over that they deserve to be where they are now. When you bring these people onto a team and tell them they are good and worthy of this chance, it works as a force multiplier on their talent. These folks have the humility of being passed over or having lost and the determination to not let it happen again on their watch.

For those whom I work with directly, I ask people to tell me what they need to do and what it's going to take to do it. We generally negotiate over their priorities or to-do list and they produce a simple work plan. Then I give them the resources and they confirm the deliverables and the timing. After that, my job is to provide support and aid to them. I course-adjust with them as needed and then try to be as supportive as possible. If a project goes astray, my job is to know it and bring it back on track or cut it if it's a clear loser. I nip at my people's heels like a sheepdog until they are back to on the right path. You have to be calm and rational about it, because if you are pushing the envelope, you are going to fail, and so we fail all the time. Failure does not bother me; not trying something worthwhile does.

My leadership style is inclusive. I want everybody to feel like they have a say in running the company. If you own something, you take better care of it than if you do not. We try to consider everybody's input. I tell our people, make me nearly obsolete as a CEO. Make me the un-CEO. Though the paradox is, someone still has to set the table for the democracy to run properly. A good CEO is like a symphony conductor: he does not make the music but makes sure it all flows together nicely.

I have to be mindful that I'm not completely "part of the gang," even if I want to be. You do not want to become everybody's friend but nobody's leader. That's a disservice. You must step up and lead, whenever necessary, and that does require a certain amount of relational distance. I must always be willing, for example, to deliver tough constructive criticism or even let someone go. But I also have to know when to back off and let people be. You can stress out an organization by banging too much on the small stuff. I made this mistake in my early days. So, I try to keep my attention consistently on the big important initiatives and give my people a break and trust them on the smaller problems. People watch where you step in. If my attention to a matter will not help our patients or front-line employees, I try not to spend time or resources on it.

That goes to how to focus as a leader. There are so many opportunities to pursue, so many problems to solve, so many people to tend to. A challenge of being CEO of a large company is remaining focused on the right priorities. Especially when you have plentiful resources, there are so many potential options, projects, directions. I believe that the wonderful payoff for being faithful and disciplined in sticking to your strategy and your core competency is, even more opportunities come to you and the better you do. Another paradox.

I feel very fortunate to be in my position. I also feel lucky that a relatively contrarian view of culture building, turning a company around, and running it has actually worked. That my original thesis of applying the Golden Rule, creating a listening organization, getting good people and giving them what they need to do their work, then making sure they know that I appreciate them, also works.

Doing 360-degree reviews *on myself* has been very helpful in informing my leadership. I ask my team and my board every year to just dump the truth on me. To tell me straight what I'm doing right and what I'm doing wrong. And I confront these brutal truths and really try to improve. I also ask for help from my team when I'm slipping or not improving. Because to be the team leader and save my place on the team, I must be sure that my team knows that I, too, must evolve to keep that place. I must know if and where I'm causing problems and what my blind spots are. I must know if and how I'm stifling debate, because I need my people to be comfortable and open with me and each other. I need to know if I am failing to recognize or acknowledge my own behaviors that are just getting in the way. I want to get strong, honest feedback, and give strong, honest feedback.

At some point, the company collectively will say, "Paul, you're gasping for air out there on the track and we need a new runner." I've given my team and board complete freedom to tell me *anything*, including when I'm holding them up and when it's time for me to go. At some point I will. Forced evolution will eventually lose to the pace of what the company requires to thrive. My hope is I will not be daft or arrogant enough to try to stay around past my sell-by date. Everybody has one. If somebody could do a better job than I can, then I want my team to have the license and the boldness to let me know to move on. It is a highly rewarding job and a wonderful experience to have the privilege to take care of people who really need us. If I could do it forever, I would. But, as Clint Eastwood said, a man has got to know his limitations.

I suppose my career could be taken as an object lesson for aspiring leaders. It has been a long road with lots of unexpected turns and still has an uncertain end point. Not everything has turned out the way I planned—by far. What's worked for me is being open, trying lots of things, pivoting when necessary, digging in where I can, and being flexible when I need to. You're going to lose—so expect it, learn from it, and bounce back. Anticipate change because it's the only constant. And always: The Golden Rule. If you do that, you'll be fine, and will most likely thrive.

Paul Kusserow is the President, Chief Executive Officer and Chairman of the Board of Amedisys (AMED), a publicly traded, leading provider of home health, hospice, palliative and personal care services with over 21,000 employees in 39 states and the District of Columbia, making over 12.3 million patient visits per year. Under his leadership, the company has grown its value 9x itself and has been established as a clinical leader in industry.

Mr. Kusserow, a healthcare industry veteran with deep expertise in driving growth strategies, previously served as President and Vice Chairman of Alignment Healthcare, Inc., an integrated clinical care company focused on providing care to the Medicare population. Prior to that, he served for five years as Senior Vice President, Chief Strategy, Innovations and Corporate Development Officer of Humana, Inc., a Fortune 75 healthcare services and benefits company.

He began his healthcare career with Tenet Healthcare Corporation, one of the nation's largest investor-owned healthcare service companies, where he spent seven years, the last four as Senior Vice President of Strategy and Ventures. He began his career as a management consultant at McKinsey and Company, Inc.

Mr. Kusserow is on the board of directors of Picwell, Oak Street Health and Matrix Medical Network.

Mr. Kusserow received his Bachelor of Arts from Wesleyan University, where he was a member of Phi Beta Kappa, a Philip B. Brown Scholar and an Olin Fellow. He received his Master of Arts with honors from Oxford University, where he was a Rhodes Scholar.

Chapter 16
Relationships and Reputation: My Path to Leadership in Senior Living

Randy Richardson

After graduating with a degree in marketing and economics from Iowa State University, I was hired as a management trainee at a marketing company, where, ultimately, I was promoted to account executive and worked mainly with one client, Proctor & Gamble. I cut my teeth in consumer marketing: sampling, mailing, and especially testing new products. My job provided me an education on understanding your customers and offering them a product that they want.

A former colleague then recruited me to interview with two real estate operators who were starting a dog racing track and I said "sure" even though this was hardly in my career plans. But I figured an opportunity like this won't happen again. There are lessons in that: cultivate your relationships and be open to opportunities, even if they sound highly unusual.

The dog track did not pan out, but the owners offered me a job as a shopping center manager at their company, General Growth. They had a strong training program, thus were not concerned that I had no knowledge of the business. After a few years and some promotions, they asked me to run the company's in-house advertising agency. I was not excited about the work itself but accepted the position because it was an opportunity to expand my experience base.

Ultimately, I was promoted to oversee a $5 billion portfolio of assets, including shopping centers and other companies that we had acquired. There I learned a lot about how to merge cultures because along with the companies that we purchased came their employees, and they did not always do things our way. Getting the people side of merger integration right is high stakes and I proved adept at creating one culture from many.

In 2000, I was recruited to be CEO of what was then Classic Residence by Hyatt, a senior living company that was founded by Penny Pritzker, a member of the

R. Richardson (✉)
Vi Living, Chicago, IL, USA
e-mail: RRichardson@ViLiving.com

M. Lifschultz, E. L. Schneider (eds.), *Paths to Leadership in the Senior Living Industry*, International Perspectives on Aging 29,
https://doi.org/10.1007/978-3-030-53966-5_16

family that owned the Hyatt hotel company. I said to the recruiter, "Gee, this sounds interesting but I'm not looking for a new job. Plus, I don't know a thing about the senior living industry." But I interviewed anyway, and Penny said, "I don't care if you know this industry. You know how to manage an enterprise." She offered me the position and I accepted, and 20 years later I'm still with the company, which we rebranded in 2009 as Vi Senior Living. I later learned that Penny had spoken with many people who knew me professionally before she ever contacted me.

The only job I ever applied for was my first one. My reputation and my network got me every job since. This illustrates the power of relationships in careers. I emphasize this within our own company. Some of our key executives have come from outside the industry because we knew of their superb reputations from people we trust.

No doubt, I learned about developing relationships from my father, who deeply influenced me as a leader. He was an insurance salesman. Ultimately, he ran his firm's sales force. Insurance sales is an intently personal business where people skills are both necessary and rewarded. My father studied motivational speaking and devoured teachings about management and leadership; he handed them down to me. Growing up in a house listening to Dale Carnegie records and reading books by Steven Covey and that ilk had a lasting effect on me.

I learned from these classics that much about leadership is fundamental, but that does not mean it is easy. Above all, you must be a good communicator. But that does not mean that you just have to speak well. You also need a message to communicate. To lead an organization, you must have a vision about where it should be, where it is going. Once you know what that future state looks like, then you have to be able to articulate it in ways that people understand and are motivated by. It cannot be ambiguous, and it has to be done consistently.

Moreover, your interpersonal skills are equally important, and part and parcel of your communication skills. Leaders with high emotional intelligence are typically more successful because they can convey on a more personal level their thoughts and ideas, and people are more receptive to that approach than abstractions.

These two skills converge in the *how*, that is, one's mode and style of communicating, and I believe it is irrefutable that a leader must put a premium on doing more listening than talking. With that approach, I believe you can create an environment in which people will be comfortable challenging the status quo. Jim Collins famously said, "Good is the enemy of great." But often, good is not good enough. It's okay, but it is the enemy of great because settling for good leads to complacency. If you want to be improving as an organization, you want everybody thinking about how things can be done better, as well as to feel safe about bringing that topic to the table for appropriate discussion. When people aren't bringing problems to you, you're not making progress. That's the day you stop being a leader.

Looking back, I always wanted to be in a leadership position like I am in right now. When I was a young professional, I was very confident and had strong opinions about how things should be done. I wasn't *always* right, but I did have a point of view. That is, I tended to see ways in which the organizations that I worked within could do better. Whether the change needed was simple or complicated, I always wanted to be in a position where I could impact a company's performance. And that means leading people.

My career path illustrates my ambition. With each successive role, I sought and was given more responsibility, which enabled me to actualize my ultimate desire to achieve a leadership position where I could guide, decide, and direct people toward excellence. In my position as CEO, I see the culmination of so much that I learned in my career, which may be summed up like this: my aspiration for Vi Living is not to be the biggest, just the best. As supportively as possible, I ask our people for their best. I believe that is what most people are motivated by, to give their all to something they feel a part of and believe in. We continually try to improve every aspect of our business, and that means we always have plenty of work to do, though it is the type of mission-driven work that people in their bones *want* to do.

Mission and its sibling principles, vision and values, are where a leader's communication abilities get rigorously tested. Having a pile of inert statements on posters and in desk drawers accomplishes nothing and can even lead to organizational drift and confusion. These three principles are fundamental, and it is a leader's highest responsibility to ensure they are fused into the actual work being done. Put another way, an organization's people must be aligned in these principles, and the leader is the "aligner in chief." Bain's definition of mission and vision suit me: vision is your "desired future position" and you simply cannot move forward without a sense of where you want to arrive; mission "defines the company's business, its objectives and its approach to reach those objectives" and unless acted upon, you will have your own people working at cross-purposes. Values are the guiding beliefs and behaviors that your organization is emotionally invested in to accomplish the vision; without values, anything goes. They all work together, and they all must be expressed through your actions every day. And again, this begins at the top.

Up until I joined Vi, it had grown over the first 13 years of its life primarily through opportunistic acquisitions. The company would identify an asset it thought had value but could be improved, then acquire it. No two communities were alike. And while the quality of the physical plants were high, the service was inconsistent. Under such conditions, it is virtually impossible to have a strong brand image and an aspirational vision.

But we knew what we wanted to do and what we wanted to be: to bring hospitality services into senior living and do it better than anyone else. I was given the opportunity to frame how we would do that. We started the conversation internally by asking, if we were to start over today, what would we want to be when we grew up? What will we do different, what will we do better? As we went into constructing facilities from the ground up, we devoted as much of our resources to developing an excellent workforce as we did to developing beautiful physical communities. This new approach went hand-in-glove with developing a set of mission, vision, and values that were unmistakably clear.

I visited all our communities and spoke with hundreds of employees. I asked each one two questions: why did you come to work for us and what do you like best about your job? Almost universally, they had come to work with us because we represented an organization that was high quality, and they wanted to be identified with excellence. That reputation was tapping into the equity of the Hyatt name. This was a great head start to capitalize on, and one way we did that was to make Excellence a core value.

The answer to what you like best about your job was also almost unanimous: making a difference in somebody's life every day, doing meaningful work. To feel that way, people must be compassionate, and to reflect that this was in our company's DNA. Compassion became one of our core values.

Finally, Integrity became a core value because you cannot exist in this business without it. Your residents, and by extension, their families, move in with you for the rest of their lives, and unlike the hotel business, you can't make short-term fixes to problems by, say, offering a free night to an unsatisfied customer. We make a solemn promise that we are going to serve and take care of our residents every day.

The strength of identifying your core values is it gives you a common lexicon, a way to talk about how you want things to be done. I tell our people, when you have a tough decision to make, one of the first things to do is ask yourself, within the context of that value structure, what is the right thing to do? When done consistently and repeatedly, it reinforces those things that you feel are most important to the organization.

Perhaps an overlooked aspect of mission, vision, and values is that together these principles substantially inform your *brand*. Branding is often misunderstood as merely the look, feel, and sound of a company's name, logo, and advertising, when in fact it is the identity, or association, that is shaped and formed by the *sum of experiences* your customers have with you. When properly seen this way, branding should not be just a function for your marketing department, but a fundamental, even sacred, responsibility of a leader. This is especially true in senior living, where you must live up to your promise to serve and care for your residents. And that is accomplished through the everyday work of all the people you lead in your organization.

When I first joined Vi, it was called Classic Residence by Hyatt, having been started by Penny, whose family owned Hyatt Hotels. The name had equity and consumer recognition from its hospitality roots, but still early on, I saw the need to re-brand the company in order to focus on our true business of providing senior living and care. When Hyatt Hotels decided to go public, our need to change our name was hastened.

In going through the process of selecting our new name, we learned through extensive research that while Hyatt's name was highly recognizable, it caused the purchasing decision for only 6% of our customers; far and away, the most powerful factor was the local reputation of a community. Our name hardly mattered. But our *brand* sure did. Even though I had solid marketing training from my time working with Proctor & Gamble, I was surprised by the limited impact of the Hyatt name.

We chose the name Vi—which is a truncation of *vita*, the Latin word for life—not to change minds but to reinforce what we were already doing: offering, through our superb care and hospitality services, a vibrant lifestyle of wellness, activities, nutrition, and socializing. Vi Senior Living is not about retirement, it's about *living*, and our new name became part of a great brand not because it's catchy but because it vividly describes our purpose. I like to say that a good brand becomes "owned" by the customer. Thus, our branding was, and continues to be, both a powerful selling tool and a platform to emphasize the importance of our culture.

Now, the fact is, even in a strong culture, people make mistakes. We intend to minimize or eliminate making the *wrong* mistakes. In orientation of new employees, for example, we discuss the specific things for which you can expect to get fired. That way, there is no misunderstanding. But even with rigorous training, you will always have somebody that somehow rationalizes an action that they have taken, or they intend to take, that is not consistent with our value system. You must call them on it. It is a grim aspect of the job, but you must make an example of those situations. If you allow somebody to act in a way that's counter to your value system, the message that is sent to other employees can be quite corrosive. You do not want to be telling everybody in the company that this is what we believe, this is how we work together, and this is how we want to serve our customer, while you've got a team member in the ranks who's doing otherwise. That will confuse your people and have them think their leaders are saying one thing and doing another. That undermines your leadership authority.

The Vi way is to shine a bright light on the actions that are unacceptable but with even greater emphasis to highlight the things that our great people are doing that are productive, creative, and fun. Thus, our executive team and our managers are always looking to recognize people who exemplify the value system, whether it be through a single action or how they work within their organization. People often think about the negative aspect of a company's values, of how to deal with employees who don't honor them. But the far important action is to recognize the actions and the contributions of individuals who are personifying our value system.

When it occurred, the Great Recession was the most significant business challenge I had ever faced. It was also a test of our mission and values. In 2007 our business was strong. We had built a number of communities and they were performing well. We had two communities under construction that were scheduled to open in mid-2008, and they were 85 percent pre-sold. The very week the first one was to open, Bear Stearns failed, triggering the financial crisis. We moved in only 15 percent of our depositors; we lost 70 percent of the total pre-sales. We were financially back on our heels. Senior living has a high correlation to the overall housing market, and the mortgage crisis hit our market hard. And when your *industry* faces failure, at thc company level all bets are off.

With financial engineering, minor downsizing, and an openness to creative risk-taking, however, we steered through those three years and came out remarkably well. We never missed a debt obligation. We even made money. We never conceded our commitment to principles, like excellence and integrity, though. We did change our business model on the fly, for example, by meeting the lowered market through price reductions and concessions, but we remained true to our basic precepts.

Our conduct during the Great Recession is a good example of how important modeling is for leaders and managers. If we had cut corners on our mission, vision, and values—even for the sake of survival—ultimately that would have destroyed the organization from within. As I constantly remind my executive team, the people that work with you and for you, are *watching* you. They're looking at how you deal not only with day-to-day business, but also bigger problems and challenges, and they're learning from you whether you like it or not. Coaching our leaders and modeling upright behavior is a big part of my job. One of the stubborn facts about leadership is, it's not who you think you are; it's how others perceive you to be.

Modeling behavior also goes to one of the least sexy but most important responsibilities leaders have and that is employee retention. Indeed, it is so crucial in our industry that, for a few reasons, it is not overwrought to call it *existential*. One, there is a shortage of care workers which is worsening, so finding new ones gets more difficult and is a considerable draw on resources. Two, your residents and their families enjoy the consistency and familiarity of knowing the people who are caring for them. And three, though it may not be an *obvious* link from A to B, retention is directly related to leveraging opportunity, in that you must always be prepared for opportunity, whether planned or unexpected, and having a solid team in place at all levels is the only way. Thus, we take steps to cultivate retention at its roots.

We emphasize training and development at the management level and offer three different management and leadership programs. It is highly valued within our company to be chosen for our leadership and management tracks because it represents our commitment to helping strong employees do an even better job and helps them get prepared for greater responsibility and advancement.

We focus on middle management positions, those that have direct responsibility for supervising employees working, for example, in the dining room, in activities, in housekeeping. The training enables the development of skills that promote a cultural environment where people feel good about their job; engaged, prepared, and supported.

In addition, we encourage employees to further their education, which we facilitate through tuition reimbursement and online training. Lastly, we make sure we have a very good benefits package, and that makes us pretty "sticky" as an employer. And we pay people well.

I'm proud to say that at 20 percent we have one of the lowest turnover rates in the industry. Half of that is involuntary, meaning we are the ones that have decided to make a change. The other half is voluntary, as some leave for a better role in another workplace that we couldn't necessarily provide in our company. The key to maximizing retention is creating a high level of employee satisfaction and engagement. The number one reason people leave jobs is their manager, not pay. Pay is always an issue, but studies always demonstrate that it actually is not at the top of the list.

Leaders have another responsibility, which is especially pressing in senior living: anticipating and planning for how the industry may evolve to meet the challenge and opportunity of the growth of the Baby Boomer population. All the same statistics and data point to accelerating demand, affordability issues, and changing wants and needs of the customer. Technology is a key area to look at, because it holds the promise of helping solve these important issues. Yet tech has not yet been deployed by the industry in a widespread way. At a recent senior living innovation conference, for example, I asked a panel audience how many operators had electronic medical recordkeeping. Less than one quarter did. Adoption of fundamental technologies like this is accelerating, but, especially among smaller operators, not fast enough. Operators will get left behind and even go out of business if they don't get comfortable with basic technology. However, the paradox is that, especially at smaller companies, technology may be viewed as a pure speculative cost center that resources should not be wasted on. But inevitably those who tend to defer technology investment will eventually get caught flat-footed.

The growth in America's aging population is going to place a lot of stress on the systems that serve it, whether that is healthcare, social services, or housing. The demand for products and solutions for older adults is going to boom. At all socio-economic levels, they will be more demanding than the current older generation, simply because they have grown up with *more*. At the higher end of private pay senior living, where we operate, that will mean more choice, more amenities, and a greater emphasis on subjective attributes like purpose and fulfillment. Opportunity will abound.

But a profound challenge, both to the industry and to the aging population, will be affordability and access. Estimates are that two-thirds of Baby Boomers cannot afford private senior living. That places the burden of creative problem-solving on the industry, and whether it's through financial innovation, design, technology, and other such factors, I believe we will see more affordable products conceived and built. In addition, I expect the government to become increasingly involved in supporting senior living. Healthcare cost containment and adequate housing standards are closely linked and that will be one important incentive for the government to deepen its role in senior housing. Indeed, I believe that presidential candidates in the near future will be running on platforms of taking care of America's older adults, going beyond talk of Social Security, Medicare, and Medicaid.

How the industry, and indeed, society, responds to such challenges will depend in part on the evolving practices of leadership. While bedrock principles of leadership and management have not changed—you can learn much on leadership from reading ancient Chinese thinkers such as Confucius, Lao Tzu, and Sun Tzu—we will see some changes generationally. After World War II, business management was modeled on the military, with hierarchical structure, policies, and procedures. That approach worked less well for the more independent-minded Boomer generation. For example, Emotional Intelligence has been identified as a foremost quality for leaders; yet this concept was not even identified until the last twenty-five years. Consider, too, how Millennials are often derided in the media as lazy and selfish, yet in research, and certainly in Vi's workforce, prove to be perfectly hardworking and ambitious; moreover, they are perhaps more purpose-driven than previous generations, which bodes well for our industry if we make it attractive to these workers.

Of course, it is the COVID-19 pandemic, which originated early in 2020, that has brought the greatest challenge to leadership in the history of our industry. As the great executive coach Brian Tracy says, "The true test of leadership is how well you function in a crisis." With the ongoing risk that the coronavirus may pose, indefinitely, to the older population as well as to the caregivers who serve them, there has never been a greater crisis for our industry.

I have led through growth, mergers, acquisitions, downsizing, and the Great Recession. Each was challenging. The COVID-19 crisis is far more difficult to navigate, of course because of the mortal risk of infection but also because there is simply no reference for the unknown. As I write this, we are in it—not through it—so time still holds the verdict as to how well we as a company and I as a leader have managed the crisis.

Our ability to recognize the significance and potential danger of COVID-19 early owed much to our overall responsiveness as an organization; it is in our DNA. I called our first formal meeting with my emergency management team in early February, 2020. I sensed that a wait-and-see attitude would not suit the situation, given the torrent of fact, theory, and speculation that was coming out, whether from governments, academics, and the media. We immediately ramped up our infectious disease protocols in all of our communities. While the deep cleaning of our public spaces was already good, the protocols being used were developed to fight Noro Virus, influenza, and gastro-intestinal outbreaks. It became quickly apparent that this situation was much different, being a *novel* coronavirus. We began initiating much more extensive mitigation measures, addressed equipment needs such as PPE and sanitizer, and planned for the likelihood that we would go into quarantine mode with our residents.

We launched communications platforms for both employees and residents (and their families) to share information about the virus, including transparency on the infections within our communities. We established daily response team conference calls with our community leadership to manage through staffing issues, as some employees were afraid to come to work. We established restrictions for all visitors including family members, stopped all outside admissions to our care environments, halted resident activities, and canceled all sales events. We established a PPE procurement team and sourced and purchased masks, gowns, gloves, face masks and shoe covers and sanitization products. And we secured a surplus of food supplies in the event of possible long-term restrictions. Inevitably, we had our first COVID-19 positive test of a resident and immediately quarantined that facility. We closed our corporate office and implemented our remote work plan.

As I write this, our residents are in the third month of staying at home. As an organization, we have found a routine in how we work and communicate. We deliver meals to our residents' living units and we offer in-home activity programs virtually. Some states are beginning to lift restrictions. Residents want to get out and about, use community amenities, and have their families visit. We have re-opening plan that will be implemented on a phased, market-by-market basis. Testing will be a critical part of this process and we have managed to gain access to testing from a variety of resources with very little help from government agencies. Reopening may be more challenging than managing our resident and employee populations through stay-at-home mode. Time will tell.

Of course, crises can be great teachers, as new lessons are learned, and familiar ones are affirmed. There are several lessons that leaders should *not* learn for the first time in a crisis, however.

First, if *prior* to the crisis you have not earned the respect and commitment of your team—especially the people who do the hard work every day—you will not be equipped to lead *in* a crisis. In other words, a crisis is not the right time to learn how to show appreciation for your people. Nor is a crisis a good time to put aside such expressions of your appreciation. We take every opportunity to acknowledge the wonderful performance of our community teams and we share those stories across the company. Our residents have been incredibly complimentary of our community leadership and we share those comments broadly, too.

Second, provide your people visibility into your thinking and solicit their input; involve them. Trust that your people have many of the answers. And, crucially, reward them with compliments for their participation in managing through the crisis and say, "Thank You!"

Third, frequent communication is critical, and the more personal you can make it the better. So, Zoom conferencing has been indispensable, enabling me to personally connect with our community leaders, community staff, and corporate teams. Regular progress reports on the support efforts for our communities has been critical and has reinforced confidence that as a company we are organized and working in concert to care for our residents and keep our residents and employees safe.

In my role as a leader, I am *always* on the lookout for ways to improve my relationships with our people, because that can only improve performance and retention, and support a positive environment. The book *TouchPoints*, by Conant and Norgaard, encourages the elegant concept of daily connections with your people to learn through listening while also using those connections to advance your agenda.

A great example of this is how to encourage creativity in an organization: by asking questions. Too frequently new and original ideas are killed by a chorus of "that won't work." I like to rescue those ideas and ask the proposer follow-up questions to explain their thinking. Often good dialogue emerges, whether it supports the original idea or leads to another one. As a leader, you have an obligation to understand group dynamics and encourage contributions under a policy that no idea is too crazy to raise.

Creative thinking comes in different forms, which a leader must be aware of. Most people fall into one of two ways of thinking. The first one is sequential, going from A to B, B to C, if C then D, etc., and it is important to have people like that in your organization because often they are the doers. Doing *is* creative. The second type is what I call a simultaneous thinker, whose idea chart looks more like a spider web, with several ideas going in their head at the same time. With the latter, you don't want to close off any of their thinking until you know more, and so asking questions will help you understand whether one of those ideas can yield a successful outcome. You must also have critical thinking, because otherwise you will have a pile of ideas and nothing will get done.

Needless to say, we all have strengths and weaknesses, and it is part of a leader's job to recognize this truth and make try to make the most of it. I am especially struck by how many leaders think they know everything. Too many act toward those who work with them with the approach of a one-way street, and they place too little value on collaboration. To me, it is a fact that you can and must learn from and rely on those who work for you. Many leaders make avoidable mistakes because they lack the complete picture when they make a decision.

Another way for a leader to fill in the complete picture is to constantly be reading and learning. I look for those sources that present and analyze familiar concepts in way that turns a lightbulb on. A standout is "*The Tao of Warren Buffett*," by his daughter, Mary Buffett. Though putatively a book on investment, it also contains gems of wisdom on leadership and management. When Mr. Buffet quotes the philosopher Bertrand Russell in saying, "The chains of habit are too light to be felt until they are too heavy to be broken," that should cause any leader to ponder how their organization may be getting stuck in corrosive ways of thinking or acting.

The classic "*Good to Great*" by Jim Collins is my go-to resource to help people to start thinking about how they grow and develop as a leader. The chapter called Level Five Leadership is a research-based discussion of his study on what makes great companies and the great leaders behind them. His conclusions are that determination and humility are the key factors. As Collins describes it, more wins go to the plough horse than the show horse.

Collins also co-wrote, with Jerry Porras, the article, "Building Your Company's Vision," which was in the Harvard Business Review over twenty years ago but is enduring and invaluable. It is a template for how to identify the mission, vision, and core values of your organization. I used it when I came to Vi and review its lessons regularly.

Working on my personality traits has been a career-long project; I suspect it will never be complete. As I tell aspiring leaders all the time, *the* most important quality for a leader to possess is self-awareness. You must know your strengths and weaknesses. For example, at times I'm too expressive. People often do not need to hear a word from my mouth to know I'm disappointed; you could say I'm a very bad poker player. But over the years, I have deliberately improved this quality, so that at least I am in control of what I say, whereas thirty years ago, I might have spoken before much thought. Still, I must remain highly conscious of the impact I can have on people. People are watching you as a leader, how you react and what you do. However, when you understand how that works, you can actually influence things without saying very much, whether positive or negative.

I advise aspiring leaders to pick one of the many personality evaluation tools available and pay attention to what it ostensibly reveals about you. Such tests are not always foolproof, but chances are that much of what it suggests will resonate. Then leverage your strengths and work on your weaknesses. Close the blind spots. Over the years I have found areas to work on, and by doing this I have become a more effective leader.

As I said, the only job that I ever applied for was my first one out of college. Every other opportunity came to me because of both professional relationships and the reputation that I had developed. I did not set out deliberately to build up these two "assets" but looking back, doing so has been the key to my success. I urge young and aspiring leaders to approach this matter of your relationships and reputation quite consciously. And, I would add, it will help your development not only professionally but personally, too.

END

Randy Richardson is the President of Vi, which he joined in 2000. Under his leadership Vi has completed over $1 billion of new development and established itself as a first-class operator of senior living communities. Richardson is responsible for company leadership and strategic growth initiatives designed to enhance Vi's economic value and further the company's position as an industry leader.

Randy has an extensive background in property development, management and marketing. Prior to his position at Vi, Randy was Senior Vice President of Asset Management for General Growth Properties, where he worked for 14 years. He has served on the Board of Directors of the National Investment Center for Senior Housing and Care and was Chairman in 2012–13. He currently serves on the Board of Directors for Argentum (FKA ALFA). Along with receiving his Bachelor of Science from Iowa State University, Randy serves on the Dean's Advisory Council for ISU's School of Business. He's also a Fellow of the Culinary Institute of America and a frequent guest lecturer on topics of leadership and management in long-term care at the University of Southern California. Among chief executives recognized in the U.S., Richardson ranked 19th and received an impressive approval rating of 96% based on anonymous and voluntary reviews Vi employees shared on Glassdoor over the past year.

Chapter 17
Making Decisions Without Perfect Information: A Key to Leadership in Senior Living

John H. Cochrane III

Like many in our industry, I did not anticipate a career in senior living. I intended to be a zoning and land use attorney, and after graduating from law school, I joined a large firm in Chicago, planning to become a partner there. In the course of developing my practice, while on vacation I serendipitously met John Erickson, the pioneering founder of Erickson Retirement Communities. John described his vision of what the "third half" of life could be like, and how a retirement community should be designed and operated to accommodate people's needs and desires.

Through family experiences, I was exposed to retirement communities growing up and always found them to be depressing, even soul-killing, places. But John was executing a radically different concept, and it seized my interest. As he described a 100-acre former seminary campus near Baltimore that he planned to re-develop, I knew I wanted to be a part of the effort. I invited myself to visit his headquarters to learn more about his business, with the unstated intention of luring John to become a client of my law practice.

John elaborated his aspirational vision for a retirement community: a place of vibrancy, fulfillment, and engagement. He was impassioned and I was moved. As our meeting wrapped up, John graciously invited me on a tour. I actually declined; I knew all I needed to know just from his description and asked him for the assignment of managing his zoning requirements. He replied, "If you are going to work for me, you must see one of our communities—then you will see what I'm *really* talking about." I'm glad he was so insistent.

We visited Charlestown, a large community of independent and assisted living, and skilled nursing care. As we toured, I saw his descriptions brought to life in ways that were unimaginable to me prior to seeing it firsthand. There was a palpable

J. H. Cochrane III (✉)
Human Good, Los Angeles, CA, USA
e-mail: John.Cochrane@HumanGood.org

M. Lifschultz, E. L. Schneider (eds.), *Paths to Leadership in the Senior Living Industry*, International Perspectives on Aging 29,
https://doi.org/10.1007/978-3-030-53966-5_17

warmth and joy in the community, among both residents and workers. John knew *everyone*, and it was obvious how well liked he was. I was impressed, to say the least.

John explained that he did not need any more attorneys for hire but was always looking for talent to carry forward the mission and vision of his company at the operating level. At which point I looked at him and said, "If you have a place for me, I would love to work for you."

I went home to Chicago that night and asked my wife if it was ok that I had "volunteered" us to move to Baltimore and take a 90% salary cut to work for John. After her initial shock, my wife said, if this will be fulfilling work for you, then I am on board for the adventure. Within weeks, I had wound down my law practice and was living in John's spare bedroom. I was even commuting to work with him. That was about 30 years ago, and I have never looked back, nor ever had a bad day on the job.

In our careers, as in every aspect of life, we often face and make decisions without having all of the information that could affect the outcome—despite the ever-growing availability of data and analytics. That was the case in my decision to drop a promising law practice to move cross-country for a job as, essentially, a trainee. But I just knew that it was the right decision. Though I did not have all the information, I did know that I was deeply touched by John's vision and mission. How people make decisions in the face of uncertainty is a major determinant of their capacity for good leadership. I recommend that aspiring leaders study some of the many interesting books in this field, such as *Thinking Fast and Slow* by Daniel Kahneman, a behavioral psychologist who won the Nobel Prize for economics, and *Blink* by Malcolm Gladwell.

The premium on such decision-making is high. As a leader, one learns the paradox that, often, the further up the "food chain" you go, the more degraded the information you receive, in part because you are further away from its origin. Moreover, people often hesitate to bring the boss bad news and to tell you a story that conflicts with your predetermined and well-communicated narrative. As a leader, I'm always scanning, sifting, and sorting information. Yet I know I still may not have *all* the information relative to a certain decision. Meantime, of course, the pressures and the stakes increase with the level of one's leadership. I believe we must pay attention to three interconnected factors in these situations: intuition, preparation, and decisiveness.

First, we must learn to trust our intuition—our gut instinct, the voice in our head—because intuition comes from the deep resources of knowledge and experience that we have gathered in our lives; it is a legitimate part of our brain's processing power, and should not be dismissed as overly "emotional." Intuition has been proven to help produce good and satisfying decisions.

Second, to the extent possible you should become an expert in your field, because the more you know about a subject, the more reliable your intuition will tend to be. This may and will compensate in those situations where all the relevant data or information is not readily available.

Third, we should develop our decisiveness, to learn to make decisions, especially where there is risk due to uncertainty. You can "decide your way to great decisions," meaning decision-making can be improved with practice. As Theodore Roosevelt said about decisions, "The best thing to do is the right thing; the next best is to do the wrong thing; the worst is to do nothing." Good leadership requires the ability to make firm decisions that others will have to—and want to—follow, whether about business practices, opportunities, or personnel matters. *Learn to decide.*

Indeed, the mistake made most by young and inexperienced leaders is a failure to act. Decisiveness is a key part of leadership. While you typically cannot wait for perfect information, you also can't hold out for the "perfect opportunity." Accordingly, you cannot expect perfect results. You are often going to be acting under duress with imperfect information and sometimes unclear expectations or anticipated outcomes. Yet you still must decide; you've got to do the work, and you must get things done. You must be comfortable making decisions without adequate time, under pressure, and with imperfect information.

This has been vividly illustrated during the COVID-19 pandemic, which we have been navigating for approximately six months at the time of this writing. No one could have anticipated the abrupt changes in the conditions we operate in, the information we receive, or the decisions we must make. Yet we had to continue to execute our mission to "ensure that those we serve have every opportunity to become their best selves as they define it" while also keeping our residents and staff safe.

Ultimately, the pandemic has had, and will have, a profound effect on how our company operates. And many of the changes brought about in response to this crisis will survive and are, in an odd way, deeply positive outcomes.

First, the pandemic has caused me, our leaders, and the organization as a whole to become much more agile in decision-making. As new information was coming into the company, particularly in the early days of the pandemic, leaders needed to assimilate disparate, sometimes conflicting data into actionable insights, often overwriting or amending decisions made only days earlier. While frustrating for many, what this afforded us is the necessity of operating much faster than we had been used to in the past.

Second, we have stripped away hierarchy and devised more flexible team-based operating processes. Our response to COVID-19 brought together a diverse group of leaders from across all channels and business lines of our company in a daily crisis call (which later evolved into what is now known as our Morning Drive), where we discussed company-wide results and overnight changes, disseminated and digested new information, marshaled resources, assigned responsibilities and desired outcomes, and created special purpose small groups to address discrete issues. This flexible approach to business planning has been a huge positive and has already been applied to new initiatives as well as ongoing business operations.

Third, we are going forward with a leaner operation. As part of our early response to the pandemic, we moved quickly to adjust staffing at both the corporate and community levels. While cuts are always hard to make, we learned to pare down only to extent that was necessary, which has helped our overall effectiveness and improved efficiencies.

Finally, the pandemic has caused us to focus strategy. We have jettisoned strategies and initiatives that, while perhaps nice, were not strictly necessary. This increased focus allows us to devote more resources and attention to those matters that truly advance the work of the company.

At Erickson, I started in the executive training program, which required learning every aspect of the business by rotating through the departments, including nursing, food service, architecture, and construction. It was a holistic introduction to development and operations. Perhaps above all, I learned about relating to a very diverse workforce *and* customer base, from the housekeepers and executives working there, to the retired schoolteachers and CEOs living there.

John promoted me from that program to lead the development and operations of his second community in Baltimore. He explained that I had demonstrated strong leadership skills, including decision-making, communications, spotting talent, motivating people, and getting results. I was so humbled, and of course delighted, by his extremely supportive words. Then he added, “Though I could be wrong. And I may have to fire you in six months.” He believed in me, but also left himself an out. It was reminiscent of Ronald Reagan’s statement about the Soviet Union: trust but verify.

John Erickson’s leadership in turn helped shape me as a leader. He provided a crystal-clear vision. He sweated the details. He was passionate. He was decisive. He made tough calls. He was fun to work with. He was demanding. He could connect with anyone. And he was exceedingly customer-focused. I aspired to have these qualities and hope that I have actualized them.

No matter what stage of growth the company was at, all actions and decisions started from the premise of, what is in the best interest of, and will produce the best experience for, our customers? The large size and complexity of Erickson developments and operations did not deter that. Every decision, from selecting sites to designing facilities to programming activities, started with a deep understanding of the customers’ needs. From *that* conversation, we backed into the realities of a sound, sustainable, replicable, and scalable business model. *The* most important lesson that I took away from my time with John Erickson and his company: customers first.

In 2009, I joined what was then known as Southern California Presbyterian Homes. The organization was in a turnaround situation, with which I had no experience; Erickson had been healthy and expanding during my time there. I quickly realized that growing a successful business with a strong culture is radically different from, and considerably easier than, starting with cultural and financial deficits and effecting a turnaround. I faced the greatest challenge of my career.

The environment was fraught. The people strongly desired to do the right thing and be successful, but significant and rapid changes had to be made—in terms of people, practices, priorities—or we could have gone out of business or lost control of the company. In fact, I later learned that several of my counterparts in competing organizations had a betting pool on how long it would be before we had to hand over the keys to regulators, lawyers, or competitors.

My first threshold to cross was, how do I walk into a 50-year-old organization—one rich in history and legacy, with up-and-running properties, and committed staff—and reorient it with a vision that enables people to be enthusiastic about the future while also honestly confronting the imminent existential threat? It would require navigating substantial, gut-wrenching changes. We had to sell properties, abandon lines of business, and replace personnel. Working with our board, our executive team, and our resident leadership, we created a vision of not only what we wanted to do, but also what we needed to do to. We did not know what the future looked like, but we did know it had to be different from *this.* We developed a *case for change.*

With that vision, we executed a significant turnaround of the company. Within two years, we merged with two West Coast providers and created one of the largest not-for-profits in our space. This experience validated the need for me as a leader to promote decisive action, clarity of vision, and a deep connection to mission and purpose, which enabled people to prioritize what needed to be changed, preserved, or abandoned.

I believe our collective openness to ideas played a large role in our success. I also believe that intellectual curiosity is a key trait; I cultivate it in myself and look for it in others. Indeed, I look for people who are always scanning the horizon for information, who keep the "mental vacuum" always running. This is a good trait for anyone, but is especially critical to leaders. Do you look at what's going on outside of your own little world? Are you able to synthesize disparate pieces of information, bring them to bear on your business, and make them actionable? Are you curious about the successes and struggles of other organizations and companies? Do you learn from others' successes and failures?

As a leader, you must have genuine humility and an insistence to hear things you probably don't want to hear. Success can breed overconfidence and a certainty of your own beliefs. To sustain success, leaders need to be constantly challenged on their beliefs, even if they seem to be well-founded and based on a good track record. As Wall Street says, past success is not a guarantee of future results.

In our field, I often see failures to think strategically and to take a long-term view. We get too caught up in running our day-to-day operations, which does consist of putting out a lot of fires. As leaders we think, if I can just get through this next crisis, *then* I will think about the future. But that fails to recognize that there is going to be a fire to put out every single day! One who waits for that glorious day of perfect calm when you've got nothing else to do but think big thoughts will be waiting for Godot. Yet in an industry that is changing rapidly and faces a radically different future, there is no putting off such planning.

Indeed, one of my ongoing priorities is creating the habit among leadership people in the organization—including myself—to routinely enjoy "white space," which is time dedicated to doing nothing but thinking. The activity can take various forms, like taking long walks or meditating. The point is, even during the workday, we must take time away from the go-go-go and just think. When we view thinking as a legitimate act of work, we frequently find new ways to solve problems, strategize plans, and just *see*. This also enhance our powers of intuition and decision-making.

The matter of imperfect information applies no less to decisions about people—whom you hire, whom you fire. You cannot know everything about anyone, but you can look for, or insist on, certain qualities that you value in your leaders and managers. Among the qualities I seek: a high comfort level with accountability; the awareness of what has made one successful, and what is needed to replicate that success; the capacity to assess challenges and devise plans for addressing them; and perhaps above all, a self-awareness of one's limitations and stubborn personality traits. It is useful to recognize which important qualities can't be trained. For example, I can *introduce* people to what it means to be curious, how to scan for disparate sources of information, and to think about applying them to our business. But, if you lack an innate sense of curiosity, I probably cannot *make* you curious.

One key decision area where you *do* have all the information is your communications, especially around your mission, vision, and values. I believe we communicate them in everything we do, and as a leader, I ultimately have full responsibility for it. *Everything communicates*. Sometimes leaders can put too much stock in what we *say*, and while that does carry tremendous weight, it can never be divorced from what we *do*. Who gets promoted? Who gets hired? Who gets let go? How are decisions made? Where is capital allocated? Every decision we make is visible and communicates a message to our stakeholders, and we need to be cognizant of that. We must always be vigilant that when we get mission, vision, and values aligned, then we will have brand integrity, and that's when communication is powerful. The flipside is, when they are not aligned, then you get brand dissonance, and no matter what you say, your communication will be ineffective, undermined by your own failure to control what you can.

Leadership, though, is not a solo act. Leaders are only as effective as their teams. Of course, this starts with hiring and firing; getting the right people on the bus, and getting the wrong people off, and doing that expeditiously. Providing clear direction is critically important, though I have had to continually learn and relearn that lesson. You simply cannot provide too much clarity as to expectations, but at the same time a leader needs to apply a fairly light touch. You should exert considerable influence on whom you hire, then lead them lightly. I know I have hired well not when I can tell people what to do, but when they tell *me* what to do. I believe in hiring great people, showing them the mountain we want to climb, then letting them choose the route and the equipment. A leader should define the where, but not the how.

Another important people issue leaders in senior living face is an overall labor shortage; in the U.S. it is bad but is especially acute in our industry. To remain successful, we've got to both retain great people *and* attract a whole lot more. Above all, we must connect the work to a real sense of purpose and meaning. The good news is, in the senior living industry, we can do this convincingly. We have a great story to tell about the impact our work has on the lives of our residents, their families, the surrounding communities, and our team members. In competing for workers, we must emphasize that our work impacts lives in a powerful way that most work simply does not. We must show great appreciation that much of the work is done by people who make pretty close to the minimum wage. Gratitude goes a long way towards motivating and keeping people engaged. We must create environments that allow people to not only thrive in a current job but also to meaningfully grow according to their talents and ambitions.

The issue of retaining and recruiting dovetails with the opportunities in senior living, which are almost unimaginably large. At a minimum we will need a great and growing workforce. As we all know, the growth of the older adult population will accelerate over the next 10, 20, 30, to 40 years, and this will accompany big changes in how the "third half" of life is conceived. There is a huge market coming. Given this opportunity, the industry could reap huge benefits and add great value to society. The important caveat is, things will look different than it does today. If we keep doing the same things in the same ways, as if we will always be serving the same people, then we'll fail our customers. We must anticipate and respond to the changing needs and desires of the marketplace.

Baby Boomers are going to do to this industry what they do to all others: turn it on its head. They are not going to wait for us to respond. They're going to reject everything we have done in the past, as well as all things "senior": *senior* living, *senior* services, *senior* health, *and senior* you-name-it. It's imperative that we understand that. And it's not just a superficial rejection of the label, or the need for a new label; they're going to reject *any* label. As a Boomer myself, I can say we are not going to buy into what we see as our grandma's retirement community. If we as a company don't adjust, other companies will. From the physical design of spaces, to programmatic offerings, to having more in-home services—like tele-health and virtual activities—we must reinvent our way of doing business. We've got to fundamentally rethink and reimagine how we might deliver on the hopes, needs, and aspirations of this population. The Baby Boom generation is a 40-year demographic gift to the senior living industry; shame on us if we fail to make the most of it.

Despite this massive opportunity for the industry, we also cannot ignore the inherent challenges that this population will present, notably on the issues of financial preparedness and health conditions.

On health, Boomers tend to think we are the most fit generation that's ever lived. But the fact is, this is simply not so. We lie to ourselves and ignore the overwhelming evidence that is readily available and observable to us all. Specifically, the epidemic of obesity will make the Boomers *at least* as unhealthy as previous generations. Obesity (and its co-morbidities) poses significant stressors on our healthcare, social service, and senior living systems, in part because increasingly people will live longer while unwell.

Financially, senior living has historically served two markets, and they are polar opposites without overlap. One is at the very high end, with life plan communities that are driven by large entrance and monthly fees. For those folks, the industry has bundled expensive services—whether needed or not—and serves them pretty well. The other is the very low-income people we have served quite well, with nonprofit service coordination and affordable housing communities.

However, as an industry, and as a society, we have not met the needs of the approximately 70% in the middle: those who are not in the top 15% of the population, and so cannot afford, or perhaps do not want, a life plan community, nor in the bottom 15%, so they don't qualify for government subsidies. For those in the middle, there is "no room at the inn." Yet this 70% that is all but ignored by our industry is the fastest growing segment of the aging population; and, data and trends suggest the segment will grow to 80%. How? Because affordable housing is getting squeezed to serve only the lowest 10%, and private pay fees are going to increase and serve only the top 10%. Moreover, it is going to be 80% of a much "bigger pie" due to growth in the aging population. So, there is tremendous opportunity, but it is going to be driven by models that we are not yet seeing from the industry. As an example, I expect great demand for congregate housing, since many Boomers enjoyed the formative years they spent in college living in dorms.

These large challenges and opportunities of the Baby Boom *will* change the roles of leadership and management, though perhaps not in commensurately large ways. The core of leadership will remain the same. However, motivating a workforce that's increasingly remote, mobile, and global is going to be an important component of new leadership. We're going to have to look beyond our physical walls for how we provide services, programs, support, and assistance to people, and there is a difference between motivating a workforce that can be gathered up in the hallway and one that may located across the county and perhaps abroad.

Another shift we will see is a pressure on leaders to manage at a faster pace within fast-changing environments. For example, skilled nursing has changed more in the last five years than it did in the first 20 of my career and we should expect such acceleration not only to continue but also to affect more, if not all, aspects of our business. Operating in a more dynamic environment is going to require greater nimbleness than most leaders in this industry are used to.

The changes in the industry mean that no leader is immune from the need to keep growing. To me, growth is about having an active sense of curiosity and a willingness to learn. That means reading, observing, asking questions, challenging my own beliefs and those of others. It also involves heightening my intellectual awareness and engagement and constantly scanning the world, looking outside my own borders, beyond my own business, and trying to understand what's going on out there! Then trying to imagine how that applies to both where our field is and where our field is going. I look at what other businesses are doing, and ask myself, what are the implications and opportunities for HumanGood? How can I apply this to our work, our business?

In addition, I have used a coach consistently over the last few years—someone to talk things over with, someone who understands enough of the business to offer helpful advice but who doesn't know the business so well that they're trying to run it for me. We work on general leadership, but the person also knows me well enough to call me on my own BS, reinforce where my instincts are right, and challenge where they think my instincts are wrong. This causes me to think through what I am doing and gain a deeper understanding and a sharpened perspective on whatever challenge I'm confronting.

Not everyone has access to a coach, of course, but peer groups are also very useful, and most of us can put one together. I have one with mostly friends in the industry; we call on each other for advice or just to have a sounding board. I don't currently have a mentor, but of course I was very fortunate to have John Erickson in that role early in my career, and his impact was lasting. My board of directors also plays this part, by bringing to bear outside expertise from different areas, such as finance, nursing, social work, and more. When I come up with an idea, they're a superb resource for a knowledgeable, engaged, supportive, outsiders' perspective.

My primary advice to aspiring leaders is, look at successful companies, organizations, and institutions that you admire; those that you buy from, interact with, respect. And study their leadership. Listen to their interviews, read their writings. And look for your own lessons that can be taken from them, then apply those lessons to your own position.

Secondly, find a mentor that you respect and admire, whether in your field or not, and ask for help. We all will be amazed by how willing most people are to help. They are pleased and flattered to be asked. Often, they remember what it was like early in their careers and how helpful their own mentors were. In my experience, most people when asked are very willing to give advice and assistance.

There are three leaders that I admire, follow, and watch for inspiration: Bob Igor, until recently the CEO of Disney, because he operated so successfully in an industry that is constantly being disrupted, and I see a parallel in ours; Rose Marwari, CEO of Patagonia, because she leads a company with a strong culture and a clear corporate conscience and keeps their actions aligned with their vision and values, achieving a model for brand integrity; and Warren Buffett, who by declining to make investments in companies that he does not understand well, demonstrates discipline and humility that I deeply admire.

This applies to senior living in particular, but I believe it is also universally good advice: Get educated on the basics of the field and its markets. Connect and engage with as many people as you can from the industry, including leaders, customers, and vendors. Get out there with your mental vacuum and suck up as many perspectives as you can. Always be thinking about the problems and challenges and opportunities in this industry. Pursue your ideas, and don't be discouraged by the fact that nine out of ten might be bad or just ok. Because one of them just might be brilliant.

END

John H. Cochrane III President and CEO

As president and chief executive officer of HumanGood, John Cochrane is responsible for the strategic direction of the company, overseeing operations and serving as the primary liaison to the HumanGood Board of Directors. With the recent affiliation with Presby's Inspired Life, HumanGood now operates 21 life plan communities and 95 affordable housing communities in six states serving more than 13,000 residents. HumanGood is one of the largest nonprofit senior living providers in the country and carries a Fitch rating of A- with a stable outlook on its core California-based credit. Previously, Cochrane worked for Lifespace Communities, where he served as chief operating officer overseeing 11 retirement communities in seven states. Cochrane previously worked as a practicing attorney specializing in real estate and finance. He has his law degree from Northwestern University and a bachelor's degree in political science from Northern Illinois University.

Chapter 18
Leadership in Senior Living: From Financing to Operating

Mercedes Kerr

I started my career in real estate finance in my native Mexico before relocating to Los Angeles to pursue a Master of Real Estate Development at the University of Southern California. Since my prior work experience was in resort development, I accepted a position with the hospitality division of a large real estate consulting firm after graduating. I was later recruited by a headhunter to join a healthcare REIT, though I knew little about the asset class. There were aspects of my knowledge and experience that were transferrable, but there was also plenty of unfamiliar territory. At the time, healthcare real estate was not well understood and not often thought of as an institutional investment-quality product. I admittedly did not realize twenty years ago that I was about to start an exciting journey in a growing industry.

From the start, my work centered on senior living settings, including independent and assisted living, as well as memory care. Despite being in the industry for twenty years and seeing much expansion during that time, I believe senior living is still in its early stages. I have had the privilege of watching it evolve during my career and may have even helped to transform it in some cases. There is a tremendous sense of mission associated with the work in this industry. While I found my way into this space unexpectedly, I could not have dreamt of a better place to build a career.

I spent twenty years in the healthcare REIT space. About half of that time, I was at Welltower Inc., the largest public company that focuses on healthcare real estate, including senior living. I ultimately had the role of Executive Vice President and oversaw incredible change during my tenure. Even though it was an institutional investment firm, there was a sense of mission in Welltower's work, which allowed me to contribute to the advancement of the industry.

M. Kerr (✉)
Belmont Village Senior Living, Lakeway, TX, USA
e-mail: mkerr@belmontvillage.com

M. Lifschultz, E. L. Schneider (eds.), *Paths to Leadership in the Senior Living Industry*, International Perspectives on Aging 29,
https://doi.org/10.1007/978-3-030-53966-5_18

Moreover, Welltower partners with many senior living operators, and it has tremendous reach. In my role as a senior leader there and through the many operating partnerships that I oversaw, I had the opportunity to touch the lives of hundreds of thousands of senior housing residents and their families, even if indirectly.

The most important leadership decisions I made at Welltower were about defining strategies that drove corporate results. I looked to extend my vision and my influence by promoting the contributions of others who could help deliver those results. This leveraged their strengths, while also expanding my own reach.

Welltower's culture invited opinions. The company had an investment committee decision-making process that involved a large number of people. Many team members were present in listen-only mode, but even though they did not drive the discussion, they were witnessing how and why decisions were made. It represented a great learning opportunity and fostered an environment of inclusivity where ideas are tested, and input is solicited. This process promoted critical thinking and offered a safety net for creative thinking. Participants could test a concept or thesis and find guidance, without limiting themselves by checking in every single step of the way. I am especially proud of having developed this process during my tenure at Welltower.

It is central to leadership to define responsibility and authority. It is also important to attach accountability to it. Authority connected to accountability facilitates alignment. As a leader develops goals and objectives individually, for the business unit, and for the corporation, they can and should develop real connections among the parts. At Welltower, I would set goals every year and then cascade them down to my team. This would assure that what we did was congruent, and that there was alignment between what I was doing, what my team was expected to do, and what the company wanted to accomplish. Goals and accountability must be cohesive and measurable as they go up and down the organization. For example, take setting expectations. Whether I set them for the team, or the team set the expectations for themselves, they would have to relate directly to the corporate objective, as well as their individual objectives and my individual objectives. They had to be thoughtful and often aspirational. Setting effective goals sounds complex, and it is. Having all those parts connect is really an important form of business alignment, and ultimately is what makes you most effective. An organization can be especially powerful once its members recognize that they rely on each other for mutual success, even when they are individual contributors. The business development team at Welltower, which I led, was composed of team members who were individual contributors, but connecting all the parts of their work was critical to producing corporate-wide results.

I observed and learned that some leaders make the mistake of trying to do too many things themselves, as opposed to training and then delegating to the team. True leaders learn to bring their team along, to train them, and to rely on their contributions for success. This is an important element in growing an organization, even if the leader feels they might be able to get to the finish line faster, or do things better, on their own.

During my tenure at Welltower, I enjoyed the support and mentorship of a superb leader, its CEO Tom DeRosa. He was a role model for and a champion of my career. Under Tom's leadership, I grew from being a successful individual contributor who focused on large transactions to assuming responsibility for teams and overall relationship management. Ultimately, I became a member of the company's executive team charged with enterprise leadership. It was such a privilege to bring my experience, my network, and my knowledge to the service of a much larger group, so that they could in turn contribute their own skills to drive the business forward. It is mission-critical for aspiring leaders to find an environment and mentors that allow them to flourish.

Because Welltower has financial relationships with some of the industry's key operating companies, I was able to develop strong personal relationships with some of the industry's finest leaders, including some who have had a direct impact on my career and growth. Two were especially impactful: Brenda Bacon, founder and CEO of Brandywine Living, with whom I now serve on the board of Argentum, and Patricia Will, founder and CEO of Belmont Village Senior Living, which I joined in 2019. Both have taught me so much over the years and were especially inspiring as female founders in an industry where there has traditionally been a lack of women in leadership.

I have focused on expanding the presence of women leaders within the industry. My former colleague, Christy Stone, and I started a group at Welltower to advance women's careers at the company. We also hosted an annual women's event during the National Investment Center for Seniors Housing & Care (NIC) conference to build a business network and to present speakers on topics of interest to rising leaders. I am also involved in an initiative at Argentum to promote women in industry leadership. I believe that the future is bright for professional women in our industry, and I plan to do my part to clear the way and assist in their growth. When I took about two years away from work to care for my two children, it was not common to do so and then return to the executive ranks. It took some pluck and some luck for me to succeed, and importantly, it took the sponsorship of a few industry leaders to help me chart a way back. Today such a sabbatical is easier to integrate into one's career and I am happy to do my part to help others looking to balance their personal lives and careers.

Both Brenda and Patricia also exemplify that senior living has largely been a founder-run industry. Over the next ten or so years, there will be a transition to leadership by a new generation of successors. Many leaders are now starting to plan their successions and are building up their benches to strengthen leadership at their companies. By definition, this will bring about change in the industry.

Indeed, when I first started writing this essay, I had no idea that I would join an operating company as part of its succession plan. In 2019 Patricia Will recruited me to join Belmont Village as its President, and I was thrilled to accept. I had worked closely with Patricia for many years, including managing many transactions that built a deep relationship between Belmont Village and Welltower. Patricia is a dear, personal friend, and I always admired her organization. I did not seek out this situation, but it felt heaven-sent. It represented the opportunity of a lifetime and an offer to lead in a very purposeful and impactful way. Belmont Village Senior Living has 31 communities in eight states and Mexico City. I could have not joined a better, more exciting company and management team.

Since I am relatively new to Belmont Village, it would be premature to discuss specifically how I practice leadership there. However, I can address the key issues that the company and by extension, much of the industry collectively, will be addressing going forward, especially those bearing on leadership.

I can speak specifically, both in the present and prospectively, to how leadership bears on the COVID-19 pandemic. Six months into the pandemic, we are still learning about this disease and how it is manifested. There may be unanticipated changes coming, and those of us in this industry will have to be prepared to adapt. We began in early March 2020 by adopting the recommendations of the Centers for Disease Control and Prevention (CDC). We quickly recognized that those standards would need to be configured for use in our space, and also supplemented. We saw that we would have to exert some bold leadership to protect our residents and staff. We voluntarily banned new admissions for a period of time, restricted access to properties to only allow essential visitors, used new processes and technologies to screen staff, restricted meal service to resident's apartments in the peak of outbreaks, and modified social engagement activities. We beefed up technology to enable using Zoom and other connectivity platforms so residents could stay safely in touch with family and loved ones. We are investing in telehealth technologies to enable the delivery of healthcare to residents without having to transport them to a hospital or doctor's office. We will continue to operate under safety protocols and strict vigilance as needed to keep our population safe, and will also return to activities when possible.

The pandemic is not only a threat to lives, it is also a threat to the viability of many organizations in our industry. For example, labor costs are rising, driven by overtime and "heroes pay" for frontline caregivers, as well as other costs, such as hiring temporary staff and providing childcare stipends. Paid leave has also added significant cost, when operators continue to pay staff who have tested positive and are quarantined. In addition, we must buy PPE and testing kits, build in extra sanitation precautions, and face higher insurance costs. The fact is, many of these costs will last beyond the pandemic, though some will recede. This will depend on how the industry and society settles into a new normal. With experience, we are becoming more efficient, and we do expect to be able to reduce costs with better planning. Having said this, for many senior living organizations, these additional costs will be a threat to the margins that keep them viable. Leaders have never had to pay such close attention to the interplay of mission and margin.

As to future growth—despite COVID-19—Belmont is in a relatively strong position because we have long, close relationships with our financial partners, thus we have been able to continue funding development plans. This environment, though, has been hard on many in the industry, as many financial institutions have paused their funding until they have more visibility into the impact of this pandemic and the industry's ability to continue to attract and serve its target population. Moreover, I hope to see fewer new entrants into the market, as their risk profile will not be easy to assess, and long-term expertise will be critical to succeed. At Belmont, we are continuing our projects without interruption, which in addition to those relationships, is a testament to the company's track record.

Yet another pandemic-driven challenge facing *all* players—big, small, and in between—in our industry is the new and more stringent regulatory environment that will demand considerable resources to manage. Not only is the pace of regulatory change rapid, but there is the added complexity of meeting requirements and following guidance from various levels of government. This can be especially complicated when you are operating in multiple states, which often have differing rules. Compliance is and will remain a big issue.

Finally, both the industry broadly and companies individually will have to deal in varying degrees with the fallout of the COVID-19 media coverage. It is unclear yet whether senior living companies will be grouped with operators of nursing homes, where many tragic deaths have occurred and on which there has been ample media coverage. Speaking only for our organization, I can say that with very clear, transparent, and frequent communications with our staff, residents, and their loved ones, we believe we have been able to maintain the trust they placed in us.

In addition to "events" like COVID-19, the senior living space is affected by cycles at the macroeconomic level, as well as by trends at the micro-level, such as new market supply. Senior living is arguably a resilient real estate asset class, due to its predominantly need-driven demand characteristics. Demographic trends suggest that demand for senior living community settings will continue to grow. Not only is our population aging, but we are also living longer. Demand for the services provided by the senior living industry will grow even through changes in economic or other macro trends.

The growing need for services and the evolving nature of our consumer requires an evolution of our strategy to address operational efficiencies, and otherwise elevate the sophistication and general expertise of the operation. To build the industry's acumen and achieve its full potential, it is imperative to develop a prepared workforce. To that end, I am excited about my ongoing work with USC's Davis School of Gerontology and Marshall Business School, where we have developed a certificate program for senior living professionals. The program was designed to complement the knowledge base of a diverse set of participants, bringing financial and business perspectives to leaders who spend much of their time in operations, while providing operational insights to leaders who are focused in the financial aspects of the business. This program was designed to develop leaders for the industry, and we have seen many rising leaders in its classrooms since it started.

As the senior living industry grows, it is becoming easier to recruit professionals from other lines of business. Until recently, the industry was not known or visible enough to attract large pools of strong applicants. As professional growth opportunities have become better defined, salaries have become more competitive, and employment in the field is increasingly viewed by applicants as steady, sustainable, and rewarding, our ability to attract talent has grown. In fact, in the era of COVID-19, while some businesses are failing, many seniors housing operators have been hiring. However, because our industry requires demanding work—sometimes in hard and dangerous conditions, such as the pandemic—leaders will need to continue to elevate the value of our work and the prospects of careers in the industry.

Training, borrowing best practices from other industries, and defining career paths are all priorities in the industry. Welltower employs cross-functional training techniques so that people can rise in their careers using what I call a "jungle gym," as opposed to rising only through a one-way, narrow "ladder." The jungle gym approach—many paths to the top—allows team members to gain a wealth of experience from different disciplines, affording them a multi-faceted view of the business, before specializing in a specific area. This involves doing temporary stints in different divisions of the business, working on specific projects, and generally gaining a more rounded perspective of the business. While at Welltower, I was able to see the fruits of that approach in the success and effectiveness of the individuals who went through the rotational training program. Each of them was a high-potential professional who advanced their ability to drive company goals as a result of this effort. I would advise aspiring leaders to seek out organizations that offer this kind of multidimensional approach to leadership development and opportunity.

Staff retention is extremely important to deliver consistent results. Turnover is not only costly, but also can be jarring to the residents, who put their trust in those who care for them. Senior living competes with other entry-level employers in service industries. Our industry ought to enjoy a great advantage, as people reap great value when they sense that they are part of a larger and meaningful purpose. Thus, emphasizing the mission and purpose of the work we do in senior living can be a formidable tool in our retention efforts. Competitive salaries can go a long way in attracting and keeping staff, but a good corporate culture can improve staff retention significantly, and good staff retention typically results in better culture; it is a virtuous cycle.

The mission aspect of the work in senior living, the actual impact on people's lives, is very rewarding for our team members. We can help our population age well, which is increasingly important since we are living longer. To live longer while also living well—that is a universal aspiration. It is important to reach every individual where and how they need to be met and our team does this very purposefully each day.

Another factor that is creating favorable conditions for retention and recruitment is that the opportunities in the senior living industry are practically limitless. We are entering a period of evolution, if not revolution, that will be marked by disruption, innovation, change, and growth, and that applies to independent living, assisted living, and memory care. That makes now and the next ten, twenty, thirty years a great time to be involved.

There are identifiable and important trends afoot in senior living, some of them stemming from the advent of the "age of consumerism" in healthcare. As consumers increasingly pay out of pocket for care, healthcare is becoming defined by consumer choice. Access to information online and growing product awareness is changing the way health care services are designed and delivered, as well as how they are marketed. Because we serve an older, and by definition, sicker population, senior living is experiencing similar changes in consumer interactions.

These two trends are converging as the gradual integration of healthcare into senior living accelerates. Care models have historically been more siloed and distinct; however, successful collaborations between specialists yield better transitions of care and superior outcomes. This approach is reflective of a value-based healthcare model, where care is being offered in alternative settings, including people's homes. This evolution in care delivery has allowed medical providers to view senior living settings as places where many consumers can be reached efficiently. These collaborations between healthcare providers and senior living providers, including the use of telemedicine, can reduce unnecessary trips to the emergency room and hospitalizations and consequently reduce the cost of care, while improving outcomes.

This trend will have a tremendous impact on senior living. Before, these settings were largely unknown or misunderstood by health systems and other providers, as well as payers and even consumers. The fact that assisted living, independent living, and memory care are primarily private pay models makes them less visible in the continuum of care, despite the fact that these locations offer a concentration of consumers, as well as many operators who will be very glad to accommodate such needs by offering an extension of the care services they provide. During my tenure at Welltower, I promoted the connections between health systems and other providers and payers, with an emphasis on the role that senior living settings could play in the effective delivery of care for this complex population.

When operated well, senior living services enhance a resident's quality of life through social, cognitive and physical engagement, good nutrition, and hydration, as well as medication management. Approximately a million Americans live in senior communities, a number that is expected to double in the next ten years. Monitoring their lifestyle and healthcare can yield valuable data that can be used to enhance their experiences and improve medical outcomes during those last, most healthcare intensive years of life.

If you compare someone who is discharged after a procedure to their home, where they live alone, to discharging the same patient to a community, where they will be cared for around the clock by experts, more often than not the outcome will be superior in the community setting. Welltower and others are spending time and energy studying and quantifying this. Most importantly, providers such as Belmont Village are facilitating these synergies, as evidenced by the recent growth in partnerships between healthcare providers, payers, and senior living providers, as well as the use of telemedicine platforms.

Another factor that will transform the industry is the inclusion of the Baby Boomers in the senior living population. Generally speaking, their life experience, desires, and needs are distinct from those of the residents we have served in this industry for the past 10–15 years. There will be dramatic differences between the type of care and lifestyle we will provide for this coming generation. It will be crucial for companies to recognize these changing consumer profiles and make adjustments in everything from design to programming to location to dining. Technology will play a greater role for residents and staff. Welltower and Belmont Village are testing the boundaries with smarter buildings, communication tools, such as video conferencing during the COVID-19 quarantine, and efficiency tools intended to facilitate operations. It will be a creative challenge to identify the most productive and useful innovation tools, and quite an effort to implement them in a timely and effective way.

The Baby Boom demographic also brings the great challenge of addressing the needs of the so-called middle market—those who can't afford today's industry average private pay rates, but also do not qualify for government assistance. This is a population that a recent study by NIC estimates will include almost 15 million seniors within ten years. To devise viable living arrangements for this middle market population will require a combination of solutions involving real estate, technology, and different care and service offerings, sometimes in partnership with providers and payers. It will take hybrid approaches to create a more affordable and attractive alternative that will draw consumers. Government funding may play a role in some solutions, as seen by the partial funding of senior living programs in Illinois and other states. There are companies already serving this middle market today, but the amount of consumers in need of alternatives will be so great that there will have to be an industry-based response to address the challenge.

On a related note, the industry—and society—has work to do to help America's older adults make the best use of their extended lifespans. While society at large mostly plans for an 80 or 85-year lifespan, more of us are living longer. This dynamic can give way to second careers and more active lifestyles, but it can also mean more people will exist with care needs. Longer lifespans are a very positive change in aging, but they will result in social and economic changes, and our industry will have to be a part of the response to such changes. Aspiring leaders would be well-advised to think about these big-picture developments, for they will present challenges and opportunities ahead.

Because I work with many aspiring leaders, I am often asked how one can become an effective leader, and what I did to grow as a leader myself.

Leaders must recognize that there is always room to improve and grow. There is always much to learn. Over the course of my career, and especially once I assumed substantial responsibility in the business, I have worked with leadership coaches. I have had several over the years, as the needs of my training have evolved. The leadership requirements I am currently facing are greater in scope and scale from any I have had previously, so I continue to work with a coach. I am also incredibly fortunate to have Patricia Will as a mentor, one of the best leadership minds in the business. I am also happy to be surrounded by exceptional people at Belmont Village from whom I can learn every day. This is by design, considering that the most effective leaders are those who gather talented and capable people to reach shared objectives.

One of the biggest challenges for me as a leader is balancing innovation and performance with cultural essence. This is especially delicate when one is new to an excellent and successful organization. It is vital to keep sight of the very things that make the organization successful, when seeking ways to optimize performance, scale for future growth, and meet the demands of an evolving consumer and marketplace. Fostering innovation while still protecting the essence of a great company's culture is a worthy challenge and one I think about often, as I pay attention to the course that the leadership team is building together for our enterprise. This requires me to be humble, listen, pay attention, and learn—all of which are habits that aspiring leaders would be well-advised to develop.

Learning from others can help to accelerate one's growth. Young professionals should seek out conferences, podcasts and webinars, interesting books on subjects related—and *not* related—to their chosen careers, and identify mentors or other professionals from whom they can learn through dialogue and observation. As for mentorships, in my experience the best ones are those that come together organically, rather than being assigned. It is helpful to have multiple mentors, or to find various sources along the progression of one's career, to match your evolution and growth.

My universal advice to aspiring leaders in the field is, play to your overall strengths, then team up with others who have different, complementary strengths. Be willing to recognize your strengths and weaknesses, relative to the different aspects of this business. Try to find that area where you perform at the highest level, perhaps even at level superior to everyone else. Allow others to fill the roles where your strengths do not apply and do not be insecure about doing so; it will actually enhance your and the organization's overall success. Dedicate time and effort and persevere. Learn about and develop your emotional intelligence. Be aware of how others perceive you. Knowing this can make you more effective, as you hone your techniques. Make accountability a priority, for yourself and others. Be comfortable delegating. Work as a team and celebrate successes as a team, but also support one another in the unavoidable failures. And perhaps most importantly, have fun in what you do.

END

Ms. Kerr has served on boards of leading national and state organizations dedicated to setting policy and standards for seniors housing and care. She was recently named to California Governor Gavin Newsom's Master Plan for Aging Stakeholder Advisory Committee. Ms. Kerr began her career in real estate development in her native Mexico. She is a graduate with honors from the Universidad Autonoma de Guadalajara and earned a Master of Real Estate Development degree from the University of Southern California.

A well-recognized health care and real estate professional, Ms. Kerr has more than 20 years of experience in the seniors housing sector. Prior to joining Belmont Village in 2019, she was Executive Vice President of Welltower, Inc., where she worked for more than a decade to establish enduring healthcare partnerships to build innovative, wellness-focused care delivery. Prior to her time at Welltower, Ms. Kerr was Vice President of Acquisitions at HCP, Inc. for nearly 10 years. Before joining HCP, she was a member of the consulting division of Ernst & Young LLP.

As President of Belmont Village Senior Living, Mercedes Kerr is responsible for the performance and expansion of the company's premier portfolio, the growth and opportunity of its employees and the care and satisfaction of its residents.

Chapter 19
Leadership in Post-Acute Care

Barry Port

I always knew I wanted to pursue a career in healthcare, even as a child. I cannot pinpoint a reason why. But I can say I was drawn to the field with a sense of destiny.

Like many, I assumed that a career in healthcare meant medicine, so I made decisions that took me on the path toward becoming a physician. In high school, for example, I enrolled in a class that consisted of shadowing doctors on their rotations. At Brigham Young University, I chose a pre-med major, exercise physiology, and fulfilled the requirements to take the medical school admissions exam (MCAT) and apply to medical school. In addition, I did relevant volunteer work, which included joining a humanitarian medicine project in Ghana where I "scrubbed in" on surgeries. My dedication and hard work paid off, as I was accepted to medical school.

During this period, I also got married, and I soon found myself doubting my decision to become a doctor, despite spending my whole life preparing for it. I shared my ambivalence with my wife and asked her if she could support my deferring school while I worked instead and figured out where my heart really was. She said, do whatever feels right. That was a relief. I did not want to disappoint her.

It became evident immediately, in my gut and in my head, that this was the right decision. But now what? I still desired to end up working in healthcare. I chose to attend graduate school at Arizona State University for dual Masters degrees, in Business Administration and Hospital Administration, planning to find a satisfying career in healthcare management. While in school, I did some stints working at hospitals, expecting that it would be gratifying and stimulating. For me, it simply was not. I found the environments highly corporate and disconnected from the patients, which undermined the fundamental reason for my interest—to be around people who needed help. The fact that hospital administrators typically work in

B. Port (✉)
Ensign Services, San Juan Capistrano, CA, USA
e-mail: bport@ensignservices.net

M. Lifschultz, E. L. Schneider (eds.), *Paths to Leadership in the Senior Living Industry*, International Perspectives on Aging 29,
https://doi.org/10.1007/978-3-030-53966-5_19

separate wings away from the hospital floor was symbolically and actually wrong to me.

After graduating, I took an entirely different direction, accepting an offer to work in supply chain management for the cellular company Sprint. I stayed for three years and, in spite of some engaging work, was miserable. I was lost in a huge corporation, where I felt like I had no voice, even though I could quantify the solid results of my work. Ultimately, I just did not feel that my job was truly important.

A friend was working in long-term care, and for some time he had been urging me to consider joining the young company he worked for, Ensign. But that field was not on my career radar, so I never entertained his attempts to recruit me. However, when I started looking for a new job, I relented and visited the facility in California where he was an administrator. I spent considerable time with him discussing his work. Then I wandered the floor. Previously, I had spent little time in a skilled nursing facility, and it was an eye-opening experience that left a big impression. Observing the interactions between the caregivers and the residents felt good to me. Many of us grow up conditioned to recoil at such places, as they are often characterized as warehouses for the elderly. But this was a vibrant, wonderful *home*.

Soon after, I enthusiastically interviewed for a position as an administrator-in-training. I accepted the job, though it required me to take a substantial pay decrease as well as to live with my in-laws and sleep in their dining room for six months. I was fortunate that they lived near the facility, but it was very challenging to be separated from my immediate family and not know where I was going to end up being assigned once the training was done. But I felt that I had my career calling, so I forged ahead.

I knew that I had found the right career because I immediately loved my job. After training, I was assigned to run a campus in Arizona and did well enough there that after a couple of years I was transferred to Texas to help turn around some struggling facilities and to grow our operations in the state. Wherever I went, I loved the residents, the staff, the company's leaders. I so enjoyed my interactions that it hardly felt like work. I loved being in the business of care. Of course, I may be biased because my career in this field has been exclusively at Ensign, but I believe there is a special quality to the care we provide that makes us different from others. Since 1999, the independent operating affiliates of The Ensign Group have provided communities with compassionate, post-acute care. Our affiliated entities offer a broad spectrum of post-acute care including assisted living, skilled nursing, and rehabilitative care. We believe we exist to dignify and transform post-acute care.

How we do that has its roots in the structure of our organization, which was established by one of our founders and former CEO, Christopher Christensen, from whom I learned much of my philosophy and practice of leadership. The hallmark is a very flat organizational structure. He founded the company after serving as the COO of another skilled nursing company, where he concluded that the conventional top-down corporate approach was not an ideal system for delivering care. Because we provide a market-specific healthcare service, it is best to understand and cater to local needs and cultures. Thus, he established Ensign on the foundation of "field leadership," whereby the company's key leaders would be dispersed throughout its individual facility operations, working on the front lines, not sitting in an office building trying to climb a traditional corporate ladder. In fact, we do not even have a conventional headquarters; as CEO, I work out of what we deliberately and meaningfully call our "Service Center," because for those who work there, our role in the organization is to facilitate, not dictate. While the CEO is put at the top of any traditional org chart, at Ensign my voice is roughly equal to those of our field leaders, so I consider my role one of support, not authority.

Christopher's logic behind our structure was irrefutable: our best leaders should be leading where our residents and our caregivers are. Each facility is a highly complex standalone business with between $2 million and $20 million in annual revenue, 100–200 employees, and a diverse array of operations, including culinary, social services, activities, physical therapy, and nursing. This requires a CEO-caliber leader who can understand and accommodate the needs of the residents, the employees, and the community. Moreover, they must possess the know-how to adapt to the local healthcare market. Christopher placed greater emphasis than our industry historically had on what kind of leader that should be. He sought those who could be superb administrators and understand the dynamics of organizational culture and clinical needs.

A visionary leader, Christopher also recognized that shared ownership makes profound sense if you intend to build an enduring institution that will bless the lives of people for a long time. He saw this work as worthy and noble, and sought leaders who would be committed to one another and the organization, through ownership, in a very deep and meaningful way. So, he elevated their relationship to the organization by sharing his ownership in the company with them. I find this generous and admirable, as well as business-savvy. Over the last 22 years, he has diluted his ownership position down to the mid-single digits by distributing ownership in the organization to other employees that act like owners, such as facility leaders and those who support them. With everyone who demonstrates that type leadership, he has shared ownership. So, we have an organization with many owners, who work with great pride and accountability. Too often, we see founders do what is best for themselves and only secondarily think of others. Christopher's way proves that you can truly make a difference in the lives of your people by rewarding the right mentality *and* have a successful business.

As a result, our people treat their responsibility in a sacred way. They feel trusted that they will do their best work, and they do so. They feel like they are responsible, which they are, and they feel ownership for the decisions they make and the results that they either do or don't achieve. I always try to act on my belief that throughout our organization we have very talented people who know a lot more than I do. The last thing I want is for people to believe that their priority is to follow some top-down protocol or chain of command rather than deliver the best care possible. I want people to feel valued and have a voice, and not to be shy in using that voice to help achieve results.

To the extent that individuals are appreciated and trusted enough to really own something and are given the freedom to make choices that impact what they are responsible for, they can do truly extraordinary things. Leadership is borne out of the trust that is placed in them. So, as a leader, I believe that fostering trust of others is a quality I always need to be cultivating in myself. I have been blessed at Ensign to see many leaders made from that trust, often coming from conventionally unexpected places; for example, we love when a certified nursing assistant (CNA) rises to become a director of nursing or some other type of clinical leader in the organization.

My chief role is to try constatnly to bring our mission to life, while constantly reminding everyone why expanding that mission is so very important. I say with a smile that I just hope I'm doing nothing to damage the trajectory of this great organization and that I am somehow helping it to progress after such a healthy history. It would truly be my greatest accomplishment if I could just not screw up the great things that have been and now are being done in the role I am in now.

My attitude toward leadership was formed largely by modeling key figures in my life. In addition to Christopher, I have had the privilege to be influenced by his father, Roy, who is a towering figure in our industry and serves as Chairman Emeritus of Ensign. He is a constant source of strength and wisdom. Roy started Beverly Enterprises, which became one of the largest nursing home companies in the world. I am also influenced by many of my colleagues, who are also dear friends of mine. Our relationships transcend the office, and these people are tremendous role models for humility, kindness, drive, and determination. Above all, my father set an example as the kind of leader I aspire to be. He would always rather be in the back helping others than in the front taking credit. He has always done acts of service that no one knows about except for the person that he's serving—and even then, sometimes they wouldn't know that Dad was involved. I have always admired and tried to emulate his example. He modeled the principle that, if you are more concerned for those around you than you are for yourself, your life will be so much better. There is immense and lasting joy in doing for others.

Explicating my leadership principles and style can be a challenge because I do not think of what I do as leading but rather as *supporting*. Moreover, I am introverted by nature; I generally do not enjoy being out front, nor do I relish public speaking or drawing attention to myself. Sometimes I look at the arc of my career and think, "Why am I here?" Ultimately, I can say that I try to evade the spotlight. Instead, I spend much of my time considering the experiences of others and reflecting on how I can help improve those experiences. Especially because we all spend so much time at work, the work experience should be a blessing to people's lives, their family's lives, and the lives of those that they serve or interact with. If leaders do all they can to make sure that the work experience is positive, people will do wonderful things not only at work, but also at home, and in their communities. They will think more like an owner, with genuine concern for the organization that they're serving. They will strive to bless the lives of those that they interact with. If I can help to bring this about, then I am fulfilling my role as a leader, as I see it.

I believe that a highly valuable by-product of this view or practice of leadership is it can help avoid some of the most common, and corrosive, mistakes that leaders make. Consider self-interest, which we see from so many leaders. They have a tendency to think about what benefits them the most, and this misguided view makes them either unfocused on or dismissive of the needs of others. It is a shameful mistake to make, though one that I have made myself. For example, I have been late completing annual reviews, when folks desperately need feedback. That is a task you must make time for; neglecting it is a neglect of another's progress. It should be a high priority. When you practice support as your first principle of leadership, however, such mistakes will be rare, as you fundamentally focus on others.

Another mistake that leaders often make is failing to give others a true voice and ownership of some aspect of the business that they ought to have. I certainly have done so. Related to that is the mistake of thinking that no one is as smart or effective as we are, so we should limit their ownership. This is such a selfish and false notion. People are far more capable than we leaders give them credit for. Often what they lack is not the ability, but the opportunity to *demonstrate* their ability. Now, the corrective is not to start just handing over everything at once. You have to show people the path and guide them into understanding what ownership and success look like. But once you consciously do that, you will be astonished by what people can accomplish with even limited ongoing guidance.

Finally, and perhaps most importantly, too often we see from leaders a lack of commitment to principles that guide one's actions. Too few leaders live by the principles they and their organizations espouse. It is easy to talk of flowery core values but then not live them. I believe that we at Ensign *live* our core values, including Accountability, Passion for Learning, and Love One Another. If I am not true to those principles, then I am contributing to what I believe is our society's ongoing loss of integrity. I strive to be a leader who studiously avoids such hypocrisy. I encourage our people to try to always adhere to our own words, while also demanding it of myself. If I am true to my leadership philosophy of looking to support others first, then I am likely to be faithful to our stated principles.

Minimizing mistakes does not mean you won't have disappointments or failures. In fact, if you are trying hard, there will be both. And I have had my share. In Texas, after we had established a beachhead and were pretty successful, I drove our acquisition of about 20 buildings all at once. Operationally, however, we were not ready to integrate them. We had typically done acquisitions of one or two at a time. The portfolio, which I had been so high on, became a drag on the organization. I felt responsible, being COO at the time. As a result, I had to drop everything else to help fix a mess that I had helped to create. For several months, I had to spend most of my time there, triaging problems that we could have avoided. It was a bad use of resources, and I regard it as a great personal failure. However, it was the result of an ambitious desire for growth. Thankfully, these 20 buildings turned out to become on of our greatest successes, but at a cost that I will never forget.

It was not my only failure, to be sure. Too many times, I have allowed the wrong leader to occupy that role for longer than they should have. That can be very damaging, both to the organization and to those individuals. Christopher taught me that one of the most unsympathetic decisions a leader can make—or *not* make—is to let a talented employee remain in a position where they are not contributing to their full potential. This lets them wallow in their misery and struggle unnecessarily. Your compassionate self might be thinking that they would be devastated if you removed them from the role, but that fails to consider how devastating it is for them to continuously flounder and degrade their credibility. This is a form of self-blinding unkindness. To help them see how they can better contribute in either a different role or even outside the organization—*that* is the compassionate act. And while it may be hard to have that conversation and hard for them to understand at the time, ultimately, and not too far down the road, most will be grateful that you have spared them from remaining in a position without having success. I have tended to think, I don't want to give up on them. But looking back, I have often just lacked the courage to confront the issue and make the change that would allow them to progress on with their lives.

Along with some failures and disappointments, I can also humbly attest to some successes. I am particularly proud of the work I have done helping the organization with some turnarounds. Often, the struggles leading up to turnarounds are based on personnel. I am pretty good at getting talented leaders to reach their higher potential. I have had some success attracting these folks to messy challenges, assuring them that they will be rewarded for their efforts, and getting their trust to commit fully to the effort. I am pretty good at getting folks to sacrifice and go where the organization needs help. I am pretty good at helping others feel valued, by giving them a significant voice, and demonstrating that it is needed, and it is heard. I care deeply about giving people that voice, and I suppose I have an effective way of communicating that.

I am also proud of many of our internal endeavors. We've established a nonprofit entity in the organization called the Employee Emergency Fund, which is used to help employees who have catastrophes, like an unexpected death in the family or impact from a natural disaster. We invite all staff to contribute, and so many do. The fund is administered by some of our longstanding leaders, and they have the ability to provide relief quickly. To provide for a colleague who, say, has lost her child and cannot afford the funeral cost is a blessed thing. And it's touched the lives of thousands of people in a significant way. We've distributed millions of dollars from this fund to employees, and that stems directly from having an organization of selfless people who care deeply about those they serve with.

We also pride ourselves on being a flexible organization and one way that manifests is in our dynamic new business venture program. We have spun off two publicly traded companies from Ensign as a result of this program: Care Trust, founded in 2013, is a REIT based on our principles as an organization, not the traditional REIT culture or mindset; and Pennant Healthcare, founded in 2019, focuses on senior living communities, home health, and hospice.

The creation of new entities to pursue these additional business lines was sparked by leaders within our organization who had passionate beliefs that there were opportunities in these areas. They were both grown in such a magnificent and responsible way that we were able to spin them off, creating shareholder value and creating exciting opportunities for these leaders. We currently have another four or five entities being incubated through this program, and who knows, one day they may also be spun off as public companies. All of our leaders are expected to evolve by adding new lines of business, offering new services, and attracting new types of talent. All over the organization we have innovations developing that create both good experiences for residents and opportunities for employees to learn and grow.

While our industry has many wonderful leaders, there is also a shortage of them. Fixing this is a major challenge for us collectively. Either there are not enough people who believe they can be a leader or too few aspiring leaders are attracted to our industry. Both are problems that we must solve. It is a fallacy that only a select few are capable of leadership. Often, all it takes is someone you respect to believe in you, whether it's a parent, a teacher, a minister, or a coach. When people demonstrate belief in us, our self-confidence grows wings and soars above the limiting expectations we sometimes have for ourselves. And leadership has nothing to do with charisma, self-importance, good looks, or bluster. Many people mistakenly associate these attributes with leadership. Indeed, those who would try to rely on such superficial qualities will fare poorly in our, and almost any, organization. If you make your work about yourself at Ensign, you will not succeed. It just doesn't work, here and elsewhere. Our resident population is vulnerable, and the care we give requires a group of people who are selfless and *caring*. You need leaders who are in it both to be successful for their employees and because they want to provide good, stable, and meaningful care. We need CEO-caliber leadership in this industry, and until we attract such talent in greater numbers, the quality of our products and services, and the public's perception of them, may well suffer.

Leadership *will* change in senior living as we pursue our goal of changing the way the industry is viewed. I believe our company is in a position to help do that, though it will take a long time, and there's only so much we can do ourselves, being that we have less than 1% of the market in the country. But we are committed to it.

I believe that our industry is actually desperate for change in how we cultivate leadership. Historically, administrators in the industry have risen through the ranks despite lacking the business background or training needed to manage with a high degree of sophistication. By contrast, at Ensign we hire most of our leaders from outside the industry. We do not find sufficient management expertise within the industry, and so our primary focus in recruiting is to attract business leaders from outside and train them in our domain. No doubt, it helps to have an aptitude for healthcare and knowledge of the type of care specifically we provide, but we find that a very solid business background, including experience either owning, leading, or starting a business, is ideal. Knowing how to lead people effectively, achieving results, managing P&L, understand the changing dynamics of the marketplace—these are the qualities we need more of in the industry. At Ensign, we bring in folks from fields such as banking, education, technology, and accounting, then teach them the details and complexities of our business. It works for us.

To those who aspire to leadership in our industry, my advice on how best to prepare is straightforward: get a solid business education; learn about the various aspects of long-term care, including gerontology, psychology, and pharmacology; and study up on leadership and organizational culture. Now, I recognize that this is not as simple as it sounds, because there is currently a dearth of opportunities for such multi-disciplinary study. I believe our industry ought to be working more closely with educational institutions to create such resources, especially since senior living and care as an industry is growing due to population demographics—and that means the need for qualified leaders will increase as well.

Needless to say, the COVID-19 pandemic has been the greatest challenge to leadership in the history of our industry. As of this writing, at Ensign we have been blessed that, overall, our portfolio has not been overwhelmed by COVID-19. As of May 2020, out of our 225 Ensign-affiliated operations across 13 states, 193 operations had no confirmed cases of COVID-19. As testing continues to become more available, we expect the number of known cases to continue to rise dramatically, but we believe we are prepared to operate in the COVID-19 environment for the foreseeable future and can adapt to the challenges ahead. It has been a unique challenge that has required a lot from all of us. Each of our local leaders actively adapted to the rapidly evolving COVID-19 environment. I have been very proud of what I have seen from their response. I have witnessed them taking extraordinary actions that go beyond the call of duty in a compassionate and thoughtful way. They have spent days and nights doing all they can to protect, treat, and comfort their patients and their employees. As you might expect, similar to what the country as a whole is experiencing, the impact has varied market by market and building by building, but I am in awe of our frontline teams. I am so grateful for their heroism and dedication during this excruciating period. As an organization, the past year has proven to be some of the most challenging times we've faced, both clinically and operationally. During this unprecedented period, our ability to learn, collaborate, and adapt has been put to the test.

We live in a time of 24-hour news cycles and omnipresent social media, and at times it can be hard to understand why more praise and admiration isn't heaped upon those leaders and caregivers that are on the frontlines in post-acute and long-term care. All of them deserve our gratitude and recognition for showing up every day to give their very best to their patients. We love our partners and are honored to be associated with them.

Our organization has taken numerous actions to provide the safest possible environment for its employees, affiliated physicians, and patients, and we have been preparing for the potential and, in a few cases, the actual surge of COVID-19 patients. With the assistance of the Service Center, our people have been successful in acquiring PPE, testing solutions, and other supplies and equipment. They have implemented staff retention initiatives tailored to the unique elements of various markets. The Service Center and field resources are providing training and helping to establish clinical protocols and safety measures in an ever-changing regulatory environment.

Our view is that we need to learn from the current reality and make long-term changes in our approach to infection control. We need to prepare for the eventuality of a resurgence of the global pandemic. We are looking at converting portions of our operations into infection recovery/control units staffed with employees that have received specialized training. We will not be caught flat-footed and are carefully reevaluating our supply chain, our protocols, and our preparations for challenges in the future. Our leaders are innovative and responsive in this way and have already developed approaches that will change how we do things in the future.

The pandemic is not the only source of change we face. Our payment structure from the federal Centers for Medicare and Medicaid Services (CMS) has changed.

State reimbursement programs will see some significant changes over the course of the next few years. Medicare Advantage is replacing traditional Medicare, so Health Maintenance Organizations (HMOs) are becoming a bigger force in our industry by, for example, cutting down lengths of stay and pushing for lower admission rates. This evolution and change requires you to have an organization with leaders that have the capability to respond locally. If instead you rely upon your corporate office to drive those responses out to the field, you will be in trouble. This is where I think our model delivers great advantages. Our local leaders hold one another accountable through a peer-to-peer method that allows them to push each other's performances meaningfully and to adapt to all these changes. As discussed, they are true owners and respond accordingly. Our model may not be perfect, but we believe it is well-structured to take on the future.

Significant changes have come and will continue in the role technology plays in our industry. Companies have had to evolve from paper charting to electronic medical records, which is sensible but sounds easier than it is. We have made significant investments to enable all of our leaders to have access, with full transparency, into every bit and byte of clinical, financial, HR, and cultural data that impact our business. An administrator in Abilene, Texas can get statistics and data for a leader in North County San Diego, down to the invoice level, or see details of their clinical survey outcomes. We insist on that transparency to ensure that best practices are shared widely. Our system enables staff to not only see the data but also apply advanced analytics to determine how it impacts their business and what areas they should be focused on to improve it. Hospitals and health plans are demanding information like never before, from readmission rates and lengths of stay to use of medications and therapy outcomes. Without such transparency, you are quickly becoming a dinosaur. If you are relying on their data to tell you how to operate, hospitals and health plans will drop you. You will not be a trusted source for their patient population, because they are not going to have the confidence that you know how to measure and achieve desirable outcomes.

Add to all of this the coming growth in the older population that the Baby Boomers will bring, and you get an incredibly dynamic industry that is in great need of a new generation of leaders. That means there will be an abundance of opportunity for those who want to care for the wonderful but vulnerable population that we serve. It means that there will be immense and gratifying opportunities to join an industry that rewards compassion, innovation, service, and a good business mind. If that sounds like you, then this is the career for you.

END

Barry Port has served as Chief Executive Officer of The Ensign Group since May 2019. He has concurrently served as a member of its Board of Directors since May 2019. Mr. Port served as the Chief Operating Officer of Ensign Services, Inc., which provides consulting and management services to skilled nursing and assisted living operations, from January 2012 to May 2019. He previously served as the President of Ensign subsidiary, Keystone Care, Inc., which includes operations in Texas, from March 2006 to December 2011. Prior to 2006, he served as the CEO at the Bella Vita Health and Rehabilitation Center (formerly Desert Sky Health and Rehabilitation Center) skilled nursing and assisted living campus in Glendale, Arizona from March 2004 to March 2006. Mr. Port holds a Bachelor's Degree from Brigham Young University in Exercise Physiology, a Master's Degree in Business Administration and a Master's Degree in Health Services Administration, both from the W.P. Carey School of Business at Arizona State University.

Chapter 20
Serving Seniors Through Hospitality, Innovation and Choice

Sarabeth Hanson

Sports played an important role in my youth. I was a good athlete and attended college on a volleyball scholarship. That was my main focus, so, like many 18-year-olds, when I was told to decide my life plan and declare a major, I had not yet given it much thought. But I knew that I liked children and working as a schoolteacher sounded promising; you get to impact lives and have time off to pursue other interests. I decided to study elementary education.

My first teaching job ended in disappointment when my idealistic plan to change the world one child at a time hit a big obstacle. As much as I cared to help my students flourish, if their parents were not equally motivated, I could not succeed. One boy in particular had high potential and I asked his mother to team up with me to help him reach his goals, but she replied, school was my business, home was hers, and they were separate matters. This ruined my plans and hopes for him. Since I had chosen this profession specifically to make a *big* difference in students' lives, I realized that it was not the right choice for me.

I interviewed for a position as a therapeutic recreation director at a nonprofit assisted living and skilled nursing community. I was not too familiar with assisted living and did not know what to expect when I met with the administrator. When I heard that the job entailed engaging the residents and keeping them busy during the day so that they could sleep well at night, I thought, I could do *a lot* better than that. So, I accepted the position.

A big part of engaging the residents meant getting to know them well—their likes, their *lives*—and I was excited about that. I adored my grandparents, but I had not taken the time to appreciate their wealth of wisdom and knowledge. I had not learned enough about how they made their way. Now I had the opportunity to do just

S. Hanson (✉)
Harbor Retirement Associates, Vero Beach, FL, USA
e-mail: shanson@hraonline.net

M. Lifschultz, E. L. Schneider (eds.), *Paths to Leadership in the Senior Living Industry*, International Perspectives on Aging 29,
https://doi.org/10.1007/978-3-030-53966-5_20

that with a whole community: to ask questions, hear stories, take in the vivid details of remarkable lives lived, accomplishments enjoyed, and sorrows suffered.

It was clear that there was little happening in the community during the evening and I wanted to change that. I went to the administrator and said, "I have a big idea. Let's produce a beautiful jazz concert under the stars. The residents will love it!" She said, "That's sweet, bless your heart. But your plan is flawed. Our residents go to bed early, it could rain, and what about mosquitos?" Undeterred, I promised a great success. "Ok, good luck," the administrator chuckled.

I planned the event. I recruited the residents *and* their families. We rented a tent in case it rained, and had it fogged for mosquitoes. I had the residents nap during the day. I did not miss a detail in addressing all the administrator's concerns. Luck was with us, as it was a beautiful starry night. Not a bug in sight. The band was excellent, the food was delicious. Everyone had a marvelous time. When the evening ended, a resident said, "Sarabeth, this was the best night of my life."

I knew in that moment, from the joy I had helped bring to them, that I was called to serve seniors. And I have been doing that ever since. I worked in that position for 5 years and I have continued working in the senior housing industry for more than two decades.

While at that job, I became interested in nursing. I would follow the nurses around and ask what they were doing and why. Around the same time my husband and I started a family, so for five years I was mainly at home raising our children, but I also studied nursing and obtained a degree. Eventually I started to feel that my identity was either "my husband's wife" or "my children's mom," and I decided to re-enter the workforce before I lost my sense of self. I knew I could be a good mother *and* help make a difference in peoples' lives, and I was hired in a nursing position in a seniors community.

Soon after, I was made director of the nursing department. Before long I realized that our company was contracting out home health services for many residents. Their wellbeing was ultimately my responsibility, yet we did not have the ability to track the efficacy of the home health care being provided. I recognized this as an opportunity for the company, so I wrote a business plan showing that we could improve the management of the care *and* harvest substantial revenues. Our CEO supported my proposal. As a result of bringing those home health services in-house, we billed nearly $2 million just in the first year. When the company was preparing to open a flagship community in Jupiter, Florida, they asked me to be its executive director.

It was a great opportunity and a great challenge for me, as I had full operational responsibility. We leased up quickly and had a major success. I enjoyed my leadership position, helping to make a difference in the lives of our residents and our associates. Then the company acquired four additional communities, and I was promoted to manage them. That included integrating our culture into their operations, always a critical but tricky job in an acquisition. Occupancy was a struggle, but we were able to move four of the five communities to 100% within a year. I was building a track record of success and this led the company to make me regional director of operations for Florida and Arizona. When I left that company, I was responsible for the operations and the results of 42 communities.

During my 14 years there, I met CEO's of other senior living companies, some of whom were recruiting me. I had no interest in leaving my company because I felt upwardly mobile, but since I always learned something and expanded my network in these meetings, I attended gladly. Then I met the CEO of Harbor Retirement Associates, Tim Smick. He was different, in that he was mission-minded in both his business and his life. We thought alike, and he was looking for a chief operating officer. I was so motivated by Tim's leadership that I took the next interview, with the executive team. I had a good feeling about the opportunity. And, indeed, the HRA mission resonated through all of his team members. I knew it was the right place for me, so I took the leap of faith and joined HRA, moving into the C-suite for the first time. I have since become president and CEO, and HRA is no longer a small player. We had 17 communities when I joined; we have 35 now, with around twenty more in various stages of planning and development.

While some people think that you're only a leader when you are proclaimed *the* leader, I do not see it that way. Leadership is a human experience. You are not a leader because your title says so, nor even because you have been appointed so. You are a leader because people choose to follow you; because you strike a chord that inspires them to act towards a common goal with you. You are able to touch and direct them on a deeper level, such that they feel they are contributing to that common goal. As Lao Tzu said, "A leader is best when people barely know he exists; when his work is done and his aim fulfilled, they will say: we did it ourselves."

Looking back to my earlier memories of peer interactions, I see that I frequently assumed a leadership role. It was just intuitive for me. As a child playing Legos with my friends I was always directing the construction, telling the other kids, "You do this, you do that." It is fitting that similar "play" scenarios are used in corporate teambuilding exercises, and there is always one person who emerges as the leader of a group. Then, playing competitive sports, I was usually captain of the team. When I was a lifeguard in high school, I was appointed head lifeguard. The pattern is unmistakable. I did not recognize its significance then, but I was a natural problem solver, able to give direction, create a vision, empower others to join me, and assign roles to make it happen. It was not due to a book I read or a course I took. I just saw leadership being practiced and modeled, and it felt right to me to assume that role when the opportunity was presented.

My first role model for leadership was my father, who was just a "natural." I am child number five of six, so there was quite a crew to lead. He was calm, kind-hearted, hard-working, and an efficient communicator. Robust storytelling was not his approach, yet you always knew what the goal was. He was a master strategist. He was chief design engineer on the Blackhawk helicopter, which was a massive undertaking with so many moving parts. He had a humble style, always giving credit to those around him. At home, he gave my mother all the credit for everything. It was always somebody else who had done something well. But I would observe and think, no, *you* were the one making this great. Early on in my career, I asked my father why he didn't take credit, which at the time *I* wanted to do on a project for my boss. And he said, “The results *are* the credit.” I would be a fraction of the leader that I am today without him.

In senior housing, I have been blessed to have many mentors. Two in particular stand out. Bill Kaplan was the founder and CEO of Senior Lifestyle Corporation, where he greenlit my home health venture, which was a key event in my career. He was a pioneer in seeing that our elders deserve to be served in an elegant way, and that left a lasting impact on me. And he taught me to take time to personally mentor others. Tim Smick taught me the profound power of culture within a company. He developed the enduring core values of respect, attentiveness, integrity, stewardship, and excellence, which are still served daily at our company, reminding us why we are doing what we’re doing.

I was fortunate to have learned the industry from the ground up. My first two positions were in middle management, so I kept a keen watch on the leaders I served and was able to learn both from great and deficient ones. One boss was an absolute dictator and that was demotivating. One boss was a delegator and that was very engaging. Thus, I learned firsthand that you must return authority to get the most from your own. Doing so makes those around you stronger, and great leaders need great teams. When a leader trusts, it makes people want to do even more than you expect them to do.

To borrow a term from systems thinking, “reinforcing loops” can be applied to leadership. The actions taken by leaders produce results, which in turn influence more of the same actions. When the actions and results are desirable, this is a virtuous cycle. Thus, great leaders will beget great leaders by identifying those with leadership qualities and giving them responsibility and upward mobility. We operationalize this concept in succession planning, whereby we require aspiring leaders to specify who would get their job if they receive a promotion. Thus, to have a path forward, you must find a successor who is as good as, or better than, you are. This system will stock an organization with strong leaders who in turn have a strong bench behind them. It will promote a culture of identifying talent, mentoring, and building relationships of trust among staff. Delegation therefore is an excellent promoter and indictor of professional maturity. I am probably most proud in my leadership to have hired, inspired, and mentored many folks now in high level positions throughout our industry. Reinforcing loops certainly played a role in this.

However, I do not believe that everyone is meant to be a leader. People inherit qualities and traits that simply make them better suited for leadership than others. I also believe that even those who have inherent leadership traits may not be *destined* to be good leaders. Good leaders must constantly work to refine, and even re-invent, their leadership style depending on, among other factors, who they are leading, to where, and why. For example, as the Baby Boomer workforce slowly retires and is replaced by Millennials, leaders will have to learn how to motivate these two very different generations. In particular, I believe leaders will need especially high emotional intelligence to lead Millennials because, in general, they want to be treated as whole persons not mere workers.

So, leadership is not a static state that, once reached, is permanent. To be effective one’s style has to evolve because so many factors around it change, including the times, circumstances, and people. Indeed, critical to being a lifelong leader is the capacity to adjust one's eye for finding the best in people under changing conditions and leading them to be even better than they were. If that sounds like the Boy Scout's credo to "leave the campground better than you found it," that's because I have adopted it as part of my personal philosophy and practice as a leader.

An underappreciated aspect of leadership is the importance of self-examination and self-awareness. It surprises some people to hear that leadership is as much about knowing and growing *yourself* as it is knowing and growing others. Even though leadership may be intuitive, you still must cultivate your leadership skills, and that must begin with an openness to self-evaluation. You must be willing to acknowledge your own weaknesses and failures as readily as your strengths and successes. You must look at your message and your delivery of it and recognize if it is or isn't being received as you intended. I ask myself frequently: How do I get everybody to understand the goal? Am I motivating them to achieve it? *Is this working?* I like to think this helps me achieve clarity in my communications.

As a result, I believe that our associates are far happier and more fulfilled than the industry average because they understand our culture, our direction, and why decisions are being made. This creates a relatively predictable environment, which simply makes people *comfortable*. They work hard, and they leave the residents and the community "better than they found them." That supports both individual and team feelings of accomplishment and fosters an ethic of care and service. It also decreases turnover and makes the company more profitable.

In our industry, leaders must be able to promote enthusiasm for work that for most team members is quite hard and equally unglamorous. I believe that personally knowing and understanding as many of your associates as possible is key. You cannot be hands-off, issuing edicts from a distant HQ. I spend time with our people, showing that I know not just their name but that their kid plays soccer—and showing that I care about their satisfaction on the job. By really knowing our people, I elevate my emotional intelligence and stay connected to what's important to them. And if I stay connected to what's important to them, they are far more likely to stay connected to what's important to the residents. You could call this part of leading the "soft side," as opposed to what you focus on when assessing operations or talking with investors: the "hard numbers."

Where the soft side and the hard numbers come to life every day is through our business strategy, which personifies my philosophy of senior living and, I believe, distinguishes my leadership from others in the industry. And that is, "I am in the hospitality business serving seniors." Whether I'm chatting with the person next to me on a plane, or speaking at a conference, that is my "elevator pitch." This is meant not only to reframe the classic image of a stinky, depressing nursing home and change people's first impression; it also accurately describes our business philosophy.

We lead with life enrichment programs that anyone would want to do, not birthday parties, bingo, and the boring basics. Whether it's cooking classes with real chefs, wine pairings, or visits to cultural and historic centers, we do far more than just provide care. To be sure, we do that exceptionally well, as do many in our industry. What separates HRA from others is our *focus* on the hospitality; we aspire to a much grander life for our residents. Our goal is to learn what each person's "best day" means to them and to make that "best day" possible all the time.

We serve two prototypical people in our business: the 53-year-old daughter who promised her parents she would never put them in a place for seniors and the 84-year-old parents who told their daughter never to put them there. The guilt-ridden daughter looks at a senior community, about to make an expensive purchase she never wanted to make, and knows she next must inform her parent that she's making a decision *for* them when they always made the decisions for themselves. This is a painful task. Moreover, they have conflicting priorities: the daughter's top three priorities are safety, safety, safety; the parent is in the fight of their life for independence. But through hospitality and abundant choice, our resident gets safety *and* independence. I believe that I am leading in our industry with this approach.

Now, leaders do make mistakes, and I am no exception. I focus intently on hiring with what I call a servant's heart, but you cannot expect to get that 100% right. Hiring is not an exact science and if you've hired a dozen people, chances are you've made some bad calls. No doubt, my biggest mistakes have been hiring the wrong people. Bad hires cause a ripple effect, impacting the people they're caring for, their peers, and their supervisors. A culture can erode quickly if the wrong people are in the wrong places. It is problematic that I can be overly optimistic and innately trust people, because perhaps I give the benefit of the doubt too easily. So, I must be vigilant to "inspect what I expect," do it early, and be ready to pivot—which is a euphemism for "fire faster than you hire."

To me, the right hire means there is a cultural fit. It is somebody who buys into our "residents first" philosophy while also demonstrating the passion to *improve*—as an individual, as a team member. If somebody is pretty happy with status quo, that's not our person for the job. We look to cultivate creativity. I say, if we're not improving, we're losing. We love diversity. We bring community leaders into the mix. We ask, how can we do this differently and better. And by sharing ideas and working with a diverse group, automatic assumptions are challenged.

We are fortunate to be doing a lot of new developments, because that gives us many laboratories for learning new levels of excellence. When we ask for creativity, we're asking the field to help us solve things. Creating a culture of innovation is asking people, hey, how would you solve this; think about this for us. In that environment, great ideas come, and one great idea begets another great idea. So, it may sound paradoxical, but you keep excellence alive by challenging it.

Excellence is a core value for HRA. We strive for excellence in *everything*. We say that excellence is surpassing the ordinary and becoming extraordinary. We actively seek examples of it from the field; tell us a story. So, on a weekly basis we receive feedback and we get a lot of great ideas. The ones that work and stick become best practices that we share throughout the organization. And the best part of that is, we give 100% of the credit to the originator of the idea, which keeps everyone motivated.

This is just an extraordinarily exciting time for our industry. The opportunities are boundless as the largest generation in American history is entering the retirement years. The industry has been talking about it for at least 10 years and now this wave is coming soon. And the Baby Boomers will do to senior living what they helped do to many institutions and industries—shake it to its foundations. As a model, think about the coffee industry. Not long ago, coffee was coffee; you had it black, with cream, or with cream and sugar. Now we order it in five sizes, with endless flavors, and insist on foamy, skinny, or flat white. By contrast, The Greatest and Silent Generations, whom we serve now, find comfort in things being the same. But we cannot do what we did before and think that the market will accept it. As an industry we've got to rethink ourselves completely and prepare to adapt. Baby Boomers have chosen a lifestyle of wellness. They have had gym memberships all their adult lives, so they will not be content to move into a community with a couple of pieces of exercise equipment in the basement. We will have to provide tai chi, yoga, spinning, and a lap pool. And coffee 1000 different ways, and cocktails before dinner. They are *not* going to watch TV and play bingo. Even with physical limitations, they'll still want to kayak Saturday morning. The question is, how are *we* going to do it?

That's a great opportunity and a challenge at the same time, and it means if you want a dynamic and creative career helping people, you are looking at one. Leaders must take on the challenge of getting Millennials and Gen Z's to see seniors housing as a highly desirable career choice. We must convince them that it's a great use of their business skills and social skills and a great mission-oriented business in an industry that is rapidly changing.

For me, this is exciting stuff. Because we have to figure it out. How can we generally improve the image of seniors housing and improve the quality of life for those living in seniors housing, so that it will stop being a *situation* that we have to promise our parents we would never subject them to? How do we make seniors housing not merely more socially acceptable, but a lifestyle people will be eager to join? *That's* very exciting.

Fifteen years from now, the industry will look very different from today; different, and hopefully better. Part of that change of course will be technology. There is vast room for technology to blossom in senior housing. Labor accounts for more than the lion's share of the operational expenses in seniors housing. Everybody is very eager to find acceptable forms of technology that could help lower costs and ameliorate the labor shortage, which is here and getting worse. Automation could certainly be used in food services as well as in some engagement and activities, both physically and geographically; think robotics for the former and virtual reality for the latter. On the custodial side, we should see more tech applied to monitoring and charting. However, I would caution any who think that technology can wholesale replace humans: *care* requires delicate hands and human interaction. Indeed, at HRA, we are not early adopters of technology. We like to see technologies succeed elsewhere before we invest our resources, which includes not just your money but your time, as well as the strategies you give up when you decide on this or that technology to replace those people or functions.

So, change and growth are two major themes of the senior living industry. They are also themes of my own leadership, as the opportunity and the requirement to be a better leader changes by the day. If I stop thinking about what I could do to be a better leader, then I should retire. Continuing to grow as a leader means to me, above all, that I leave people better off than they were before. I hope that this philosophy does not change. I also believe in investing my own time: giving myself to future leaders, providing what I can to help them grow and prosper in their career.

Recently I went from being COO to CEO, and I got much more involved in raising capital for our development pipeline. But I could not give up inspecting what I expect from our operations. And I had to keep making sure the culture is right at the field level. One day it dawned on me that I was no longer doing the things that I had been doing my entire career—focusing on creating a thriving, attractive workplace. So, I spent the entire summer doing a "culture tour," getting out and sitting down with associates, awarding folks with our core value recognition pins, and thanking them for their service. I wanted to be sure that I was experiencing what I was expecting to experience. That was a shift from where I had been, which was mainly getting deals done. Now, I still have to do that. But I also need to make sure that I'm continuing to use the fundamentals that have led to success. Therefore, being the leader is about managing my balancing act, making sure one thing doesn't become dominant; making sure the soft side stays a big part of who we are, maintaining a high *organizational* emotional intelligence, watching the data and the numbers but also keeping our heart in it.

Thankfully, there are many paths to growth. I am active on several industry program boards, which provides a great opportunity to interact with fellow CEOs in our industry. I regularly attend a conference that is outside of our industry—in food and beverage—and that is a great experience for me, because I want to improve my performance not just in seniors housing but also in hospitality. This is another way to challenge my own assumptions about my leadership and our strategies.

When I'm asked by any young aspiring leader—in our industry or not—what can I do to advance my career, I emphatically recommend, learn to play chess! Chess players have to plan ahead. They must think strategically. They need to know the implications of their next move. Chess promotes good decision-making skills. It shows the relationship between decisions and outcomes. You learn accountability for your actions. It teaches you to mix boldness and caution. You learn consequences, as you make a move and get both immediate and long-term feedback on what happens. It's best to start playing chess young, but I see our residents learning it at 90, so it's never too late.

END

Sarabeth Hanson was appointed President and Chief Executive Officer of Harbor Retirement Associates in January of 2018, after having served as its Chief Operating Officer since September of 2012. In addition to experience as a seasoned Chief Operating Officer, she has successfully demonstrated her ability to inculcate culture and a standard of excellence in organizations resulting in portfolio, occupancy and margin growth. Early in her professional career, she knew that she was called to serve seniors and has been successfully making a difference in the lives of seniors ever since. Sarabeth has been recognized for being a servant leader and for her commitment to excellence in her ongoing pursuit of creating hospitality-oriented senior living. Hanson emphasizes that she is driven by a desire to inspire and develop others. She is a student of leadership and is always looking to better herself and her teams.

With over 25 years of experience in the senior living industry, Ms. Hanson has become an expert in refining and shaping how residents are served through innovatively designed communities and spaces. She is passionate about fresh designs that are supported by unparalleled operational standards to ensure the ultimate living experience. She spearheaded the development of a one-of-a-kind program for residents with mild cognitive impairment—HRA's signature Sound program.

Ms. Hanson currently serves as a member of the board of directors of Argentum, the nation's largest senior living association, where she serves as Treasurer and chairs its finance committee. She is also a member of American Seniors Housing Association (ASHA) Executive Committee. She has addressed the National Investment Center (NIC), Argentum, American Senior Housing Association (ASHA), Senior Housing News, InterFace Seniors Housing, and other industry associations and conferences.

Ms. Hanson is originally from Connecticut, where she attended Central Connecticut State University as a student athlete. Shortly after completing her education, she relocated to Florida, where she and her husband John raised their three children, Kylee, Julia and Jake. In her spare time, you will find Sarabeth with her now-adult children, boating or traveling.

Chapter 21
Creating a Vision for a Better Tomorrow: Thoughts from a Leader in Senior Living

Kai Hsiao

Like many leaders in senior living, I did not plan to enter this field. I studied communications at the University of Arizona, intending to go into advertising. After graduating, I took an unpaid internship at the legendary ad agency Chiat/Day, which was then at the pinnacle of its industry. Since I was not being paid, I had to live with my uncle, which entailed commuting about 3 hours per day. Some would call that sacrifice, but I was just doing what I had to do to improve my career prospects.

After leaving Chiat, I worked for several ad agencies to bulk up my experience, then founded my own shop before joining Cramer Krasselt. One of my clients was Macerich, the large retail REIT that owns, operates, and develops shopping malls. They enticed me to join the company, and I left advertising. My wife was jealous of my job because I spent all my time visiting malls.

Canyon Ranch, the renowned resort spa based in Tucson, Arizona, then recruited me to their development team. Among other projects, I got to work on Canyon Ranch - Miami Beach, a "condotel" that was the first residential community designed for wellness. Just to be working in real estate development was fun, but Canyon Ranch is a high prestige brand, so there were always celebrities and movie stars on the scene, which added some glamor to my job.

A lot of our projects were being funded by Lehman Brothers, which went bankrupt in the 2008 crash that led to the Great Recession. In fact, there is to this day some confusion over who owns the land under certain projects, because Lehman had held the debt that financed them. Around this time, I was exploring opportunities with the private equity firm Fortress Investment Group, which in 2007 had made the largest deal ever in senior living by acquiring Holiday Retirement, a leading, pioneering company in independent living.

K. Hsiao (✉)
Eclipse Senior Living, Lake Oswego, OR, USA
e-mail: kai.hsiao@eclipseseniorliving.com

M. Lifschultz, E. L. Schneider (eds.), *Paths to Leadership in the Senior Living Industry*, International Perspectives on Aging 29,
https://doi.org/10.1007/978-3-030-53966-5_21

Until Fortress recruited me to join Holiday, I had never thought of working in senior living. Before accepting a job with them, I visited a Holiday community in Phoenix. As I was walking the grounds, a resident asked me what I was doing, mistaking me for an inspector. When I told her I was considering a job with Holiday and wanted to get a feel for the place, she invited me to meet with a bunch of residents and get the "real story." This chance encounter led to my sitting and chatting with 10 residents, all of whom were members of the Red Hat Society, a social group for retired women. They went around the circle, talking about their families, their lives, and what it meant to be together and have friendships now in this community. That sealed the deal. By bonding with these women, I knew I had found my calling. We *must* have bonded—for years I exchanged Christmas cards with some of them. As I drove home, I called my wife and said, "Does this sound crazy? I want to be a part of this. Let's do this." And that's when we decided to pack up and move from Tucson, Arizona to Portland, Oregon and be a part of the Holiday turnaround story.

At Holiday, I got to wear many hats, starting off in sales and marketing, then moving into operations. I became a "fixer" for the CEO, whether it was revamping the HR department or overseeing a new venture. I ran a Nashville-based home health company that we acquired. Then I was dispatched to Shanghai to help arrange a joint venture between Fortress and Fosun, mainland China's largest conglomerate, to create a senior housing platform. Soon I opened and ran our acquisitions group.

Holiday's then-CEO, Jack Callison, was my boss, though actually I was hired two weeks before he came on board. Before joining, I had been told that we will soon have a new CEO, but not who it was. It was the only gig I've ever signed on to without interviewing with, let alone knowing the identity of, my eventual boss. I took a leap of faith because I was so excited by Holiday's prospects. You could say I had my fingers crossed that it would work out, and it did. Jack was an awesome boss, teacher, and mentor; I would not be where I am today in the senior living industry without his mentorship.

In fact, after Jack left, I got the call to be the CEO of Holiday and had the privilege to be in that position from 2012 to 2015. During this period, we closed approximately \$5.5 billion worth of transactions, marking a period of substantial growth. As I said, I had already been hooked on senior living, but due to my tenure at Holiday I became an industry lifer. I got to know so many of our residents and their families, and I learned the business inside and out. The bottom line is, senior living is a great business, because what you do actually has a very positive impact on people's lives.

By this point, I figured it would be interesting to be on the finance side of the ledger, so I joined HCP, a healthcare REIT now called Healthpeak, and focused on their senior housing platform. It was an extraordinary education, as I worked with 26 different senior living operators and got more than just a peek "behind the curtain." From that exposure, I sensed opportunity.

So, I gathered best practices from my varied experiences in senior living, hospitality, retail development, and advertising, and in late 2017 founded Eclipse, which manages independent living, assisted living, and memory care communities for ownership groups across the United States. Within sixty days of launching, we took over the management of about eighty communities and as of this writing in mid-2020, we manage around 110 communities. We have enjoyed quite robust growth for a startup that was founded less than three years ago. Our main clients are among the biggest names in the business, including the REITs Healthpeak and Ventas, as well as the private equity firm Apollo.

Looking back on my career, I can see clearly that my parents were extraordinary, indispensable role models. Early in their marriage, while my mother was pregnant with me, my father proposed that they move from Taiwan to the States for the possibility of a better life. He moved here to seek work before I was born, and he was hired by Florida State University; the first time I saw my dad was when I arrived at the Tallahassee airport. It was so brave of him to make this move, and it was equally brave for my mother to support his idea, which meant raising me by herself in our native Taiwan for a time. Moreover, this was Florida in the '70s and '80s, and being Asian we certainly had to adjust to some significant cultural differences. But both my parents knew that the sacrifices they were making by leaving their home, family, and friends in Taiwan would ultimately pay off for their kids and them. I am also so proud that my mother, who had been a stay-at-home mom, eventually studied computer programming at the University of Arizona and worked for IBM. Their story represents a part of the great tapestry of immigrant success stories. I'm certain that without their modeling of courage, optimism, and a fierce work ethic, I would not have achieved as much success as I have.

They taught me my chief principle of leadership: creating a vision for a better future. They did this consistently through their actions. At Eclipse, we have adopted the Japanese concept of *kaizen*, which means "change for the better" or "continuous improvement." This is fundamental to me. My greatest responsibility as a leader is to define end goals that are aimed at positive change, whether for our residents' lives, our staff's experience, or our bottom line. Then I rally our people to support such goals and chart out the directions for getting there, all while communicating the vision clearly and constantly.

I also believe that there is nothing like modeling excellence in performance and getting results. I learned this aspect of leadership by synthesizing various lessons from key experiences. In the advertising business, I learned that if you don't perform, your client fires you. At Macerich, I learned how to manage multi-site operations, which is not something that would come naturally to anyone; complexity must be taught and observed. At Canyon Ranch, I learned the how-to and importance of building consensus; it's the only way to get big things done. At Holiday, I learned the power of growth and scale, and how to achieve them.

Thus, I can condense my leadership principles into this: present a clear vision so that everyone knows what the goals are, what the end result is, and what success will look like—then help drive the team to achieve results. Without such leadership, people will get demotivated and complacent, and your organization will drift or decay. Of course, this process always relies on effective communications.

Getting results is critical. As a third-party management company, if we don't perform, our clients will simply replace us with another operator. But results-oriented leadership should not be seen fearfully by those in an organization. A results orientation creates opportunities. Great results distinguish you from your competition. They validate your efforts. Ultimately, they define your brand.

Indeed, I believe it was achieving results that enabled me to be approached by some great companies offering wonderful opportunities. In business, people come to know you by reputation and by your results. This is especially important in an industry like senior living where there is a relatively small number of players at the top. A lot of us know one another. That was key to our success in starting up Eclipse. People knew that we were able to achieve results at Holiday while under a bright spotlight, as it was a very public turnaround situation. As we were launching Eclipse, it was standard operating procedure to call potential clients—and to get calls from them—and have outcome-oriented conversations about working together.

Many on the team at Eclipse worked together at Holiday, including our chief operating, administrative, and investment officers. We understand our respective strengths and weaknesses, our roles and responsibilities—when to stay within our lanes and when to merge. One learns as the years go on that you can't do it all. When we're young, we often think we can do everything. Then you mature and suddenly realize, what a fool I must have been. These days, I look to bring in people who can actually do their job better than I ever could, which helps explain why we have such a great team.

I have not always been the easiest guy to get along with. I have tended to drive hard and impose on others the expectation that they will keep up, even though not everyone subscribes to that style. My attitude was, if you want to be a part of it, great; if not, move on. So that can be tough on certain people. Over time, I have softened up a bit and operationalized the notion that folks run at different capacities. Being adaptive in my management style generally yields better outcomes. You've got to be able to manage different perspectives and know that your point of view is not always the right one.

I also believe this approach helps us avoid the pitfalls of groupthink, which is endemic in organizations. It is important to note that consensus decisions are not the same as groupthink. Consensus is desirable—when everyone agrees with a decision after all material disagreements are resolved. Groupthink is when the desire to agree prevents individuals from presenting alternatives and challenging impending decisions. Everyone just wants to get along.

As stated in the Harvard Business Review essay *Making Dumb Groups Smarter*, by Cass Sunstein and Reid Hastie, groupthink causes companies to "bet on products that are doomed to fail, miss out on spectacular opportunities, pursue unsuccessful competitive strategies." Groupthink is especially common where you have either highly cohesive teams or strong-willed leaders. We have both. Because I can be naturally intimidating, I deliberately cultivate a rapport with people, so they won't feel shy about contradicting me or trying out bold ideas. When everyone on my leadership team is leaning in one direction, I often take the opposite point of view in order to expose the contrast, offer an alternative, and stimulate people's feedback. In fact, in meetings we often assign a contrarian just to oppose the group's thinking. I believe this can smoke out everything from dormant great ideas to tension in the ranks. It also gives people cover to speak up who otherwise might have thought, this is not a hill I want to die on. I make the wry observation about my team that, on the plus side, we have been together for a while; on the minus side, we have been together for a while.

That can lead to a stale dynamic. I believe it is incumbent upon leaders to foment challenges to their own thinking, as well as to the thinking of those on their team.

Accommodating dissension, frankness, and bold thinking also opens the paths to creativity and innovation, which is a big priority of my leadership. It starts with keeping an open mind. As noted, our attitude and approach is based on the Japanese philosophy, *kaizen*—constant improvement. At the community level, we solicit from residents and staff any ideas that might improve operations. At the executive level, we do regular retreats where we talk about where we are today and what we need to do better. I say always, even if it's not broken, let's break it and see if we can do it better. By the very nature of individuality, we all have different points of view, and I want to ensure that they get aired out. We have very robust conversations. This approach also results in better plans for execution. Results-oriented debate, or even conflict, is a good exercise that I believe yields stronger mental "muscles."

A muscle that a leader should not want to strengthen is their ego. Humility, especially when you are in the business of care, should be deliberately practiced. Encouraging people to challenge you, as discussed above, is one wise habit to have. Perhaps most importantly, a leader must get out of the office and see what is happening on the front lines. I visit at least two of our communities every week; it's in my DNA. I learned as a developer that you must frequently be on site and in the field. From HQ, everything seems perfect. Things are seldom perfect. You won't know what's broken unless you walk the site, the grounds, the halls. You won't know the truth until you're out there.

When I visit communities, there is no pomp and circumstance. I do unannounced visits, in jeans and a golf shirt. I'll sit around in the community and wait for someone to notice. I always spend time with residents, and take in a sense of how they're doing, and I always observe how associates are interacting with residents and one another. I also deploy my personal trade secret: I gather intel from the hairstylists—they know everything! Often that person has the longest tenure in the community, so they can provide not just a snapshot of how things are going but a historical perspective, too. I joke that one of the reasons why my hair is so short is because I spend so much time in our hair salons. It is a surprisingly rich source of information, because the resident clients share so much—the good, the bad, and the ugly—with their hairstylist in a community.

Another favorite tactic is to ask residents and staff, if you were the boss for a day, what would you work on? What's on your wish list of things to do? If I were to give you $100,000 to spend on this community, what would you do. I try to convey to people that I'm there to help them achieve their goals. It takes some humility from a leader to recognize that there are others who have the answers. Doing two or three visits every week allows me to get a darn accurate reading of our organization's vital signs, and that really informs my leadership.

II see an intangible quality to my leadership that is related to vision and to competence and to *kaizen*: the ability to "see around corners," or see inflection points coming. This is not some unique or mystical sense, though it sounds that way. It actually comes from rigorously and continuously scanning the horizon for shifts in your industry, other industries, and society, then preparing your organization not only to survive such changes but thrive from them. Here is how I apply that to leadership.

Future leaders in senior living will have to be incredibly sophisticated, with knowledge of healthcare, regulatory policy, hospitality, finance, technology, and more. So, if you're an aspiring leader, you better know your stuff. Today, many leaders come from either real estate or healthcare. In the future, each will be a necessary but not sufficient expertise to have.

First, as we live longer, acuity will increase amongst residents. Moreover, insurance companies will increasingly channel customers to senior living as alternatives to hospitals, simply because our environments could be equally therapeutic but cheaper. That means senior living will blur with healthcare in far more ways that it does currently. As a result, our industry—already complex and highly regulated—will face more stringent regulations.

Second, the industry is twenty years behind hospitality and ten years behind multifamily housing in terms of life cycle. Many executives in senior living started off as developers and thus have a real estate-oriented point of view. When they're looking at growing their business or building the next community, they think of it from a real estate perspective—utilizing square footage to maximize ROI. But increasingly people are attracted to senior living not just as dwellers of the real estate, but for less material factors like social life and style of care. Those factors are getting more attention. We will be measured by many for the quality of outcomes we provide. These metrics will be, do we provide a rich life, not just do we have a cafe. I think there's going to be an evolution in this regard. Perhaps it will be a revolution.

Third, leaders in our industry will have to understand how to invest in and manage technology. Currently, much of the technology being offered to our industry is driven by entrepreneurs and engineers who have spent little or no time with caregivers or at a community to actually understand what is useful and meaningful. Or it's just derivatives of products and platforms re-purposed from other sectors, like enterprise systems from skilled nursing or multi-family. Technology should be purpose-built to meet needs. We see a lot of solutions in search of a problem. Future leaders will have to get their technology providers to help us care for residents, enhance staff capabilities, manage our operations, and improve our bottom line, such as a truly integrated enterprise system. Moreover, you will need to get the timing right between anticipating where technology is going but not adopting it too early and getting stuck with a needless cost. Looking at this issue today, a lot of folks in this industry have jumped on the Alexa bandwagon. Residents interfacing with Alexa is a great concept, and I do think people are doing well with that. But I've been waiting to see how HIPAA affects this, given that these virtual assistants can leak personal data and information.

Fourth, there is a huge opportunity coming in the middle market, which is projected to number 14.4 million older Americans by 2029. The group is defined as those who don't qualify for government housing assistance based on their income yet cannot afford private-pay senior living at today's rates. This is not only an opportunity; it may become an industry obligation. Where will people go if they need housing and services but don't check either of those boxes? The threshold question for those looking to enter this market is, how will we provide the necessary bundle of services in a livable environment that will make it financially profitable while also affordable? That may well require hybrid models that rely on design, technology, financial engineering, public subsidies, and more. Even if insurance companies take the lead in this market, as I believe they might, there will be opportunity and challenge for those in our industry to step up with entirely re-imagined approaches. I have invested considerable time studying this issue, including working with the National Investment Center for Seniors Housing and Care (NIC) and its brilliant, indefatigable founder, Bob Kramer. At Eclipse, two of our brands, Embark and Elmcroft, do serve this market. Our services and amenities may not be as high-end as others, but our prices aren't as high, either. Still, there will be a lot of work to do in this area for companies and their leaders.

One inflection point *no one* saw coming was the pandemic. There was no COVID-19 "master class" for senior living executives. As the virus spread, and the threat to life became concentrated in the older population and certain types of congregate living facilities, many protocols had to be developed on the fly. We assiduously gathered information from the Centers for Disease Control, the Centers for Medicare & Medicaid Services, State and local authorities, and even sources in other countries, such as Starcastle Senior Living in China, since they already had experience with the virus. We also kept a close eye on what was happening at skilled nursing facilities. We used all this information to quickly develop best practices and institute them throughout the company.

Above all, of course, we endeavored to keep our residents and our staff safe. In addition to implementing safety measures, which happened on the frontlines, leadership had to constantly be communicating developments with clarity and transparency. We are speaking with an older, often vulnerable, population and the change in residents' routines was terribly disruptive. We had to address concern and confusion around events; "What is happening?" Then we needed to explain the need for new restrictions; "Why can't I see my grandchildren on Mother's Day?" Later it became justifying the continued restrictions; families asked, "I can go back to work but I still can't visit my mom?" Again, transparency in communications ruled the day. Simultaneously, we had to address the impacts on associates. It was up to leadership to acknowledge associates are in the line of fire, and to honor them for their work and sacrifice. We had to work organization-wide to teach proper use of personal protective equipment (PPE). Frequently we encountered staff struggling to cope with limiting their personal interactions outside of work, even with their own families, as it may increase dangers to residents. We have been so moved by the dedication of our workforce.

The pandemic raises longer term issues for the industry. My strong sense is, only the strong will survive. Larger companies have the people resources to address needs, such as monitoring multiple state regulations and reporting, which are evolving daily. They have the financial resources to manage the unexpected costs, such as PPE and "Hero Pay," or bonuses for frontline workers. As in other industries, there will be fewer senior living companies when we come out of the pandemic. Indeed, we are already hearing about operators being unable to pay their leases, breaching loan covenants, and other negative financial impacts.

Unfortunately for the industry generally, senior living has been forgotten in terms of priority for PPE, testing, and CARES Funding. It has been a double hit, because while we are not receiving the assistance from public institutions that we deserved, the media collectively has taken little trouble to distinguish between senior living communities and skilled nursing facilities, where so many of the nation's deaths have, tragically, occurred. The industry will feel the effects. Going forward, the penetration rate for senior living will likely flatten or decrease, though we will still achieve growth due to the growing demographic.

This is not meant in any way to discourage aspiring leaders from entering, or remaining in, our industry. As always, in challenges there are opportunities. Plus, senior living fundamentally will always be a business where you can help people and have a rewarding career. For aspiring leaders, I advise that whenever you can, reach out and learn from others. I have learned so much just from talking with those great folks who have been around for a while and paved the way for this industry, such as Loren Shook, Brenda Bacon, and Patricia Will. If you're looking to enter the senior living industry, network your way to finding a great mentor. Getting involved with trade associations like NIC and Argentum is key, for learning and networking. Be a continual student of the industry. Reach out and talk to peers. My talks with fellow CEOs in the business are always so helpful. I love hearing what other people are doing and how they're tackling the same issues that we have. Also be a student of adjacent industries. I still attend multi-family conferences and retail conferences, because there are just so many new practices and advances being done elsewhere that could be applied to senior living.

Learn, listen, and look ahead. And keep in mind the concept of *keizan*, a wonderful basis on which to build a career—continuous improvement, change for the better.

END

Kai Hsiao is Chief Executive Officer of Eclipse Senior Living

Kai was previously Senior Managing Director—Senior Housing Properties at Healthpeak Properties, Inc. (NYSE: PEAK) a fully integrated real estate investment trust (REIT) that invests primarily in real estate serving the healthcare industry.

Prior to Healthpeak, Kai served as President and Chief Executive Officer of Holiday Retirement, the second largest senior housing operator in the United States, growing the company's occupancy, NOI and EBITDA through organic growth and acquisitions, recapitalizing the company through triple net and third-party management transactions, and centralizing operationally intensive functions. During his tenure at Holiday, Kai held several leadership positions, including overseeing operations, field support functions, acquisitions and new business strategy. Previous to Holiday, Kai worked at Macerich (NYSE: MAC), a retail REIT, where he worked in operations, marketing, and both ground up developments and redevelopments. And prior to Macerich, Kai worked at world-renown spa resort, Canyon Ranch, where he focused on new hospitality and real estate developments.

Kai serves as the Vice Chairman on the Argentum Board of Directors, is on the Executive Board of the American Seniors Housing Association (ASHA) and the NIC Board of Directors. He graduated with a Bachelor of Arts in Communications from the University of Arizona.

Eclipse Senior Living is a national 3rd party manager of Independent, Assisted and Memory Care communities across the United States under various brands, including Elmcroft, Embark and Evoke.

Chapter 22
The Paradox of Leadership in Aging Services: Confronting Harsh Realities with Incredible Optimism

Sean Kelly

My brother and I, having grown up in a "picture-perfect" family, were on the brink of adolescence when our parents divorced. We were buoyed by our parents in a different way and encouraged to take advantage of the many supports available in our extended family. I especially gravitated to my grandparents, Nana and Bobby, whose home was always filled with friends and relatives from all over; it was like a clubhouse, lots of in and out, lots to eat and drink, always a story or ten, lots of laughter, palpable caring. I remember my grandmother had the ladies over on Thursdays to get their hair styled and trade gossip over cocktails. The chatter was loud so as to overcome the noise of their hairdryers, which back then looked like a spaceship around your head. I was lucky to be welcomed to join in sometimes, and while everyone else had their "hi-balls," my grandmother, Nana, made me the most beautifully bizarre concoction from Coke and orange-juice. It was a signal that I was together with them and it made me feel just at home and free to join in. In the same house I also got to meet all sorts of other folks from different walks of life—and often of much older age—as my grandfather, who was the town doctor, would have friends, family and even patients, who otherwise had nowhere else to live, stay with them. Some of these old souls became my friends, too. Looking back, they were dearer to me than I realized.

Those experiences—hanging out with older people and doing what seemed like completely normal stuff—is part of why I am who I am. I know now that these experiences helped me to have an ease with older people, a genuine curiosity around their stories, and a respect that helps explain why I cherish my work in senior living.

After graduating from college, I was lucky to get a job with a construction management firm that was working on a huge re-development project in Camden, New Jersey, while I was also taking (and re-taking) courses which I needed to apply to

S. Kelly (✉)
Kendal Corporation, Kennett Square, PA, USA
e-mail: SKelly@kcorp.kendal.org

M. Lifschultz, E. L. Schneider (eds.), *Paths to Leadership in the Senior Living Industry*, International Perspectives on Aging 29,
https://doi.org/10.1007/978-3-030-53966-5_22

medical school. Although I wasn't certain that was the path for me, or that I had the chops to see it through, it seemed right at the time. In my job, I was the assistant to the assistant to the assistant superintendent—if the boss needed coffee, I got it; if someone needed a count of the number of plumbers on site, I counted them. I was assigned to a very dynamic construction site, with hundreds of people working on a giant project, and I was eager to learn how it all worked.

After a while on the job, I noticed that about once a month, a mysterious group of mostly men would show up in dark cars, dark suits, and dark glasses. I was curious—who *are* these guys? Of course, they were the "developers" of this giant project that I had the teeniest part in. Further, I learned that their work entailed so many facets, including construction, finance, design, marke ing, operations, economics, workforce development, even politics. The more I learned, the more I thought pretty seriously, *that's* what I want to do. It sounded like conducting an orchestra. In my mind, even though I couldn't play a lick of music myself, I might play some part in helping make the music happen, finding common purpose with a wildly diverse, dynamic group of people, under pressure and doing something that might make a difference. That would be the beautiful music. I gave up the dream of medical school, focused in a new way on the work I was lucky enough to be doing, and enrolled in a couple of other classes to supplement my knowledge of business and accounting.

In the mid-90s, I got a job with a small construction and development company when commercial real estate was slumping. But it turned out something called assisted living was burgeoning. Instead of opportunities to build the next suburban office campus, we got a number of requests to do increasingly large and complex senior living projects, and I became the point person to identify new relationships and puzzle out various financial structures for our executives to consider.

By circumstance, I sensed that I was finding a niche in senior living, so I attended industry conferences to network and learn, and I read widely to grow my knowledge base. For example, Ken Dychtwald's book *Age Wave* was eye-opening, as it projected how the needs of aging Baby Boomers would change the world, and affect my industry, thirty years hence. I also started to develop my own fundamental views on the field and its customers that I hold to this day: first, respect the people that you intend to serve, listen to your market, and don't assume that *you* know everything; second, understand the power in community and the incredible desires that people have for connection and social engagement; third, promote greater community mobility and integration—don't assume that residents of senior housing of any sort, from life plan communities to assisted living, lack the ability and desire to circulate through the rest of the world and be a part of it.

At a conference in Atlantic City, I was lucky to meet Avery Rockefeller and Doug Powell, who ultimately invited me to join their company, Retirement Living Services (RLS), which was based in Connecticut. At the time, this seemed like another world, given that I'd lived my whole life in the "Philadelphia part" of southern New Jersey. Avery took me to visit the first project they'd developed, Riverwoods in Exeter, New Hampshire. Despite my own personal experiences and as much as I was learning about the field of senior living, at that time I still carried the bias with me about what to expect in a retirement community. While I expected something dull and sad, when I walked into the great room at Riverwoods, late on a Tuesday afternoon, I was blown away by the energy, the engagement, the smiles, the laughter, the welcoming vibrations. I had to overcome the shock of my own expectations not being met as I thought to myself, this would be loud and vibrant for a high school cafeteria, even though it was filled with people mostly in their *eighties.* I had a visceral response that this is the kind of place that my very social grandmother, Nana, who was then in the throes of decline, would have enjoyed. A place where she would have thrived. And it just inhabited me: a sense that this is the kind of place I want to be a part of creating for people that I care about. It was an earth-shaking moment that caused me to say an emphatic "yes!" to a job offer despite having to pull up stakes and move to Connecticut. Fortunately, my wife Susan was very supportive of uprooting ourselves from the friendly confines of the neighborhood where we grew up, met, and lived—soon to be with our first baby.

I was with RLS for 10 plus years and learned what one would expect to—about development, transactions, finance, marketing, operations, and design. But perhaps more importantly, I also picked up many key lessons in soft skills, like listening, communicating, leadership, and emotional intelligence. For example, Avery taught me, directly, the importance of leaders delegating, and what a gift it is when you give people the latitude to "own" a situation, but also remain nearby as an inconspicuous safety net in case they mess up. The careful faith Avery put in me made me work harder, and be more thoughtful, more curious, more engaged. Avery Rockefeller took a chance on me—actually, many chances—that I'm still not sure I deserved, but remain ever grateful for.

As I got more involved in the operations that followed our development projects, I began to think, maybe that's where I belong. On a visit to one of our recently opened projects around Philadelphia, The Hill at Whitemarsh, I walked into the pub and saw one of the residents who I had gotten to know and grown so fond of over the many years before the place opened. She had been widowed and thus had considered not moving in, but in this moment, she was aglow, managing to have the time of her life, playing cards and laughing with friends. Seeing this one moment of exuberance opened me up to the idea of moving back home and trying life on the provider side by joining Kendal, which was then recruiting me to help re-imagine their development business and the potential growth of the organization.

Kendal is a not-for-profit organization which includes communities, programs, and services for older adults across the country. Kendal has done important work leading the discussion that has resulted in the almost universal removal of the use of restraints on older people across the field of senior living and associated healthcare environments. Consequently, Kendal is also known for its advocacy and empowerment of older adults determined to achieve their full potential, on their terms. Too often society sees the aging process as declinist and limiting. Kendal sees the realities that come with age, while at the same time celebrating the possibilities that go along with it.

The idea of Kendal began in the late 1960s, when a committee of Quakers set out to find and create "a better way in retirement" based on Quaker values. The first Kendal community opened in 1973. Since then, we have affiliated with 13 communities in 8 states, with another in development in California. Our communities have either been developed as Kendal affiliates from the ground up or were existing communities that have chosen to become a Kendal affiliate. Kendal may be characterized as operating under a "federated" model; the Kendal organizations are each independently incorporated but are stitched together by a set of bylaws and an affiliation agreement. There is strength in the independence that is in effect at every Kendal, but also in the association that the Kendals are a part of: one family, connected by agreements and working in common with a commitment to see Kendal's aspirational values present and alive in our culture. The Kendal Corporation is designed to support its operating affiliates and also to drive growth so that the organization may expand on its mission and have greater impact on more people determined to discover what is possible as we all age.

John Diffey, who was the president and CEO of Kendal until 2016, is an icon to me. He brought an incredible blend of heart and head that I hope this sector finds more of. He took his work to a mission level. From the time that I came to Kendal, despite my own focus on development, John gave me access to Kendal's boards and operations, and he taught me so much about our work and more importantly why it matters. In part because of John's mentorship and the graciousness of so many colleagues, when it came time for John to retire, it was neither unlikely nor absurd to most that I might be an adequate successor. I've been with Kendal for about 10 years, and I have been President and CEO of The Kendal Corporation since 2016.

As of 2020, we are expanding. This includes doing new projects, expansions, and re-positioning at nearly all the existing Kendals. And we are quite deliberately raising the bar for what we will do on ground-up projects. For example, we are proudly working with the University of Pennsylvania, where some retiring alumni and emeriti faculty invited us to help envision an intergenerational residence meant to re-define community. We sought and found true partnership with the impressive innovators at Penn Medicine and Penn Geriatrics, and we are hoping to create not just a residence but the quintessential center for geriatric health and well-being in the world, a place where older people will go for education, clinical care, primary care, socialization, health, fitness, and community.

Kendal has also entered a unique collaboration, after a 10-year conversation, with the San Francisco Zen Center, to develop Enso Village, a Life Plan Community in Healdsburg, California. We discovered that we and the Zen culture share so much, perhaps above all how we approach issues with an open-mindedness about evolution and the possibility for continuous improvement across all areas of work and life. We also share an affinity for the quiet. Having found ourselves doing and thinking along similar lines but merely using different nomenclature, it felt natural to become business and spiritual partners.

Indeed, it was easy to co-create a vision of mindful aging, taking in such elements as the joys of nature, environmental stewardship, and contemplative care. For this community, we are exploring new possibilities in senior living, such as having a working farm and a working restaurant available to the public on the property, with residents doing some of the work. And we're integrating healthcare, not by way of a standard clinic, but through a center of health and well-being that's proactive in its approach, not reactive. Most importantly, these ideas are being brought to us by the people that will actually live in this community. As I write, we have over 1000 people on a priority reservation list. There has been incredibly strong validation of the concept and the partnership.

Though neither yet exists, in both the Penn and Zen projects we are being in the truest sense mindful of, and responsive to, what the world is asking for now, not what it *used* to be asking for. We are creating energy centers that are designed for the older people who will live there, but they're also designed to be accessed by those in and around the neighborhoods in which they exist. They're designed to be *porous*, meaning they will have spaces through which people will flow. People will surely not mistake these places for conventional commercial or residential environments. We are bringing to life our rejection of the notion that institutionalization is good enough for older folks.

In the Quaker faith (as in others), there resides the idea of "continuous revelation," and my secular translation of it is, as good as something might be in the moment, there's always a chance to make it better, and we are better people for constantly seeking to accomplish that. I like to believe that this applies to the work that senior living is trying to do collectively as a sector, that Kendal is trying to do as a company, and that I am trying to do as a person and as a leader.

I trace my recognition of continuous revelation in leadership to my wedding day. At that point, I never imagined that someday I'd be the CEO of any organization, let alone be seen by anyone as a leader. When my wife and I married, she was a 4th grade teacher. After the ceremony, there was a receiving line. It was dark and cold, with snow on the ground. Yet standing there to greet us were about 20 kids—and many of their parents—from her school class, which was in a town almost an hour away. Note, they were not invited. But they wanted to be there because their teacher inspired them. She took her job to mean not just teaching the kids, but also knowing and honoring them.

What I learned in that moment was a greater obligation to be authentically present with others. To this day, that has informed how I try to meet people where they are and to appreciate them in ways that help them to come along. It was in that moment that I recognized the impact you can have on people by fully showing up for them. It also primed me to receive the extraordinary influence of others.

In addition to Avery Rockefeller and John Diffey, I have been inspired as a leader by Bob Kramer, founder of the National Investment Center for Seniors Housing &

Care. Bob has for decades passionately worked to expand and improve our field, and he continues to find powerful ways to humanize aging. Freeman Hrabowski, the president of University of Maryland, Baltimore County—a center of excellence in STEM education and performance, particularly for minority students—inspires me simply by his continued presence on the national scene. And there's Bruce Springsteen, with his four-hour, leave-everything-on-the-stage rock-n-roll shows. What I take from them all is that they are always on the move, thinking about and doing the next thing, always seeing what once was and improving on it to bring what comes next. And they are all over 70-years-old. In addition, there are people that I get to work with on any given day who inspire and teach me through the care with which they do their work and their search to always do it a little better.

I hope that my leadership has also been informed by the work of my friend Sigale Barsade, a professor at Penn and an expert in emotional intelligence (EI). Her research is about reading, understanding, and managing the emotions of others and yourself. She believes that exercises can raise EI, much the way athletes can train to improve their performance. Her work shows that we can change our brains and our behaviors to be better at EI and thus, as we navigate the people in our lives, leave ourselves and others a little better off. I do hope that the people around me can sense that I work on my EI and that they experience my gains, however incremental. I recommend her work to aspiring leaders.

Especially in our sector, where the ultimate product is *care*, empathy in leadership matters. As a result of practicing EI, I try hard to stay aware, catch myself, notice body language, take in tone. Perhaps I will catch a grimace in a meeting, which may lead me to rephrase a statement or pose a new question. This causes me to connect with people better, and I think that helps me lead better. People appreciate when you're paying enough attention to notice an averted eye or a shifting in the chair; that goes a long way not just in the moment but in the relationships that leadership is so heavily reliant on.

Another Penn professor, Adam Grant, has also shaped my leadership. The big idea in his life-changing book *Give and Take* is that you should be ready and quick to give what you can to help others—introductions, advice, mentoring, *whatever*. Give yourself, fully. What a powerful notion. Be generous in spirit, generous with time, generous with accolades and credit. Just be generous. I aspire to be a giver, and I hope that I conduct myself accordingly.

When I consider how I can do better in giving to those in our organization, I often think of a stereotypical drill sergeant barking orders at young recruits and commanding those troops to sing call-and-response military cadences as they do their grueling runs in boot camp. Inevitably, the company breaks into people at the front leading the charge, a pack in the middle, and the folks bringing up the rear. And that sergeant—at least twenty years older than the recruits—runs back and forth, goosing them up, coaching them along, inspiring them to keep pace and get through it together. I believe that is a fine example of what leadership needs to be doing constantly. The sergeant is helping each individual appreciate more in themself than they ever have before and is also motivating the group to perform well as a company. That is leadership communications aligning with leadership in action, and that is what I am always trying to do.

As the president and CEO, I must keep in mind that it is natural for everyone at Kendal to look to me to lead, but I in turn look to redistribute leadership to them. We are all a part of a kit of parts, and all of those parts are different, and all of those parts are necessary. To put the kit to good use does require from me a vision and a destination, then clear directions for how we might get there. But my approach or philosophy relies on *shared leadership.* My leadership asks for, if not insists upon, engagement. It consists of everybody's different perspectives coming together. To me, it is fundamental that folks are, and feel, invested in their work and empowered to carry it out. To give them a say in leadership is essential as a practical matter, and I also believe it is a way to demonstrate a fundamental respect of everyone I work with.

I promote a culture that aspires to see the good in every person and it is optimized when we are synergized, working together, and listening to one another. The culture is based on values that first intend to respect and honor something in every person. Everybody contributes something to it. And we actively notice that we all contribute to what we're doing right now and to what might be. Again, that's akin to the Quaker notion of continuous revelation—no matter how well things are going right here and now, they could be even better tomorrow. And by coming at every situation with a mind that is open—a beginner's mind as my Zen friends would describe it—we can constantly explore the possibility of discovering something a little bit better.

When discussing EI, I mentioned noticing. I aim for a *culture* of noticing: paying attention, holding things up to the light, speaking to things that are going well and not so well, and having the inclination to notice and find things in the small moments that are extraordinary because they teach so much so well.

I was touring the skilled nursing neighborhood in one of our communities where an esteemed man was living. He was impaired; he could not do well on his own. He was eating lunch with the help of a staff member. A visiting colleague who was mentored by this man many years ago asked me if he could say hello. I relayed his request to the administrator, who discreetly informed the aide kneeling beside the resident; he was struggling to take in a spoonful of soup. The aide explained that an old friend wanted to see him, and finally he was able to muster a response—he tilted his head toward her with half an eyebrow raised in consent. She noticed his response. She allowed him to choose. Many others might have denied him that choice because he appeared to be disengaged. With a simple question, she gave him agency and humanity.

Her gracious act was connected to a larger state of affairs; it is a perfect example of person-centered care. This is what it means to know and respect and honor the whole person and every person. That moment captured our organizational values, our aspirational culture. That was also a profound act of leadership, as the staffers took the time to notice and care and to lead by doing. Yes, I love to find extraordinary lessons in seemingly ordinary moments.

Recently my father-in-law was battling an illness that he ultimately succumbed to, and I had the sacred opportunity to deliver care to him, such as helping turn him over in bed. This connected me like never before to the work so many of our people do, and as a result, I decided to go through our CNA (certified nursing assistant) program. I wanted to do this for two reasons. One, it is mission-critical work that I want to understand better. And two, I want to be able to walk the talk and do a better job of helping people understand, from multiple perspectives, why this work matters so much.

With humility, I believe doing the program speaks to authenticity. Not that every leader in senior living has to have hands-on experience. But I have never been a caregiver. I have not led an operating community, except in brief transitions. I have not worked in our communities. But I know there is a deeper understanding to be gained of the essence of our caregiving and of our caregivers. It is also a chance to lead by example.

To lead by example, you must have integrity, because modeling behavior that lacks integrity will quickly degrade an organization. My first project in the senior living sector was at Piper Shores in Scarborough, Maine. It was very complicated, and ultimately a remarkable success. However, it did not always seem destined for that outcome. A lot of factors contributed to its success, but one of them happened to be my ability to bring people together in a way that they all felt heard. There were many parties at odds with us: neighbors sued to change the project; the municipality demanded we pay twice the fair property tax; the builder wanted more money; the nonprofit ownership naively messed with the business terms; the guys throwing hammers in the middle of the winter wanted time off; the marketing staff anguished over what delays would do to sales. Meantime, the bankers and the investors were betting on the leadership—namely, me—to get it all done right.

What served me well was authenticity, such that when I said I was going to do something, people believed me. And when I said that I would *try*, they understood that that was different from a promise or a guarantee. I was careful not to over-promise and under-deliver. Despite not having all the knowledge, experience—and certainly not the gravitas—I managed to keep the parties aligned with us. I can say, with due modesty, that I kept my integrity. And twenty plus years later, I believe I can say I still have my integrity very much intact. There is, in my view, no more important quality in a leader. Period.

Sometimes with even the best intentions, and full integrity, as a leader you make mistakes. Recently, I wanted to stimulate our aspirational culture and find ways to appreciate our people, but not with gift cards and choice parking spaces. We devised the Better Ways Project, a software platform that invited the entire Kendal community of 8000—residents, staff, family members, board members—to note individuals' good works, ideas for innovation, best practices, pilot projects, you name it. The effort required more structural muscle than we had. Awareness and buy-in was high, but the efficacy of the system lagged. We had started the project full bore but scaled it back to roll out one community at a time. It was a misallocation of resources. I know that I'm prone to making this type of mistake—moving too fast because I assume that everybody is as ready as I am and that I understand where their dynamic priorities lie. It is so important to have your team not only eager but also *ready*. It's got to be both. I sometimes mistake people's excitement about an idea as evidence of their being eager and ready. I've got to do a better job of being more aware of the difference.

However, failures, disappointments, mid-course corrections, abandoned projects—dealing with all of these are almost *de rigueur* if you are trying to lead boldly toward making things better. Leadership invites, if not demands, that vulnerability. Leadership requires a strong willingness to be wrong, the ability to recognize when you're wrong, and the humility to proclaim it. I'm really fortunate that I'm so well-practiced at being wrong.

There is little that makes me feel more engaged in work than to witness a colleague—or myself!—start out with a hard position, then reconsider the facts, the conditions, the arguments of others, their own reference points and declare, wait, I hadn't thought of it this way; maybe we could try to do XYZ differently than what I just said an hour ago. Changing one's mind represents so much of what I believe leadership requires: humility, attention to others, being present, being open to what's possible, and of course, integrity.

Leadership in senior living must continue to improve, and for almost too many reasons to list. Here are a few: The booming opportunity ahead. The competition. The primacy of mission. The demands of a changing market. The intersection with a rapidly changing healthcare system. Add to this the dearth of exceptional people to do the work in the first place. Leadership in this field needs to be capable in business and communications, as well as innovation and change management. A leader must be adept at stimulating growth, scaling operations, pushing costs down, and freeing up resources to meet an ever-more demanding population of residents and consumers who are going to be in need of more and more services.

I see three distinctive qualities that today's and tomorrow's leaders need to bring to their work. We used to be very satisfied with operators who could drive quality and operational efficiency. If we were lucky, that person could also develop and deploy strategic thinking. If a miracle occurred, they could influence and inspire a culture, too. Going forward, I feel we will need triple-threat leaders who can do it all. We need those with the strength of character to lead the organization and attract others who are ready and eager to join the mission—that is, to influence and inspire a culture. We need keen operators. And we need ace strategists. If you have a leader lacking in any one of these three qualities, as an organization you will be less than what you need to be—less than what the market will demand, less than what the staff will demand. You will not be the organization your residents and their families expect. You will be ripe for being overtaken by more sophisticated, more capable, more culturally aligned organizations moving into this space. And this is happening.

There are steps that we in this field can take. We've got to stop being the runt of professional pursuits. We should be moving and shaking, getting major institutions to recognize our importance as a sector. We should be partnering with insurance companies. We should be creating our own insurance companies. We should be knocking on the doors of university presidents to make novel deals, not just being grateful when they let some of our residents take a class. We should not only have the chops, but we should have the confidence and the gravitas to collaborate with other sectors. Kendal is doing that now with technology, higher education, and healthcare. My inspiration for believing that we should and can do that work at a much higher level comes from Bob Kramer. I believe that if we raise our game as an industry, we will attract better leaders.

Not just leadership will change. The field as we know it today will fundamentally change, too. To appreciate this, all one must consider are the two key factors of healthcare and access.

Our sector and the traditionally understood healthcare systems have been at or near points of intersection for the last several years. The need to work in partnership more directly with our colleagues in healthcare has never been more apparent from a clinical and health/well-being perspective and the perspective of providing good population health. Fortunately, we have leadership on both fronts eager to innovate.

Back in the day, health services were solely for the continuum of care. Now we think about healthcare and health service as more akin to healthy living. So, to us, the moment somebody moves into a Kendal is the moment that they are engaged in what it means to maintain or build their own best life. How do you stay as healthy as you can for as long as possible, as opposed to responding to a healthcare need? That's a big shift in our field. We are realizing that so much of what gets in the way of quality of life and positive healthcare outcomes is driven not only by your socioeconomic standing or your own genetics, but also by behaviors, many of which can be influenced by programs that can be delivered either in a senior living setting or in a wellness setting that can be accessed by neighboring communities.

Somewhat related is the issue of access. Roughly between 1% and 10% of the age- and income-qualified market has given evidence over generations of wanting to live independently in a market-rate senior living community; in mature and well-educated markets that percentage, though, can peak in the 20% range. It's not just socioeconomics that is the limitation, it's also the sensibility and the market perception. It's about awareness. We've known for a long time that not everybody wants to live in these communities, even if they can afford it, in part because too few people actually appreciate the things that can be available to them in such a community, whether bricks-and-mortar or virtual.

It presents a mission opportunity for organizations, including ours, to expand reputation, brand, and business, by taking these centers for health and well-being and ensuring that they are publicized and made available to an audience that we can scale to, an audience that would include people in the greater community within some prescribed area. In order to do that, we need to have the right kind of spaces, the right kind of staffing, and programming around activities like mindfulness. We need to have social programs that help drive outcomes and health and well-being in the population. I believe we can, and must, demonstrate that the senior living communities of tomorrow will be places where a greater portion of older adults will want to live, not have to live.

Put simply, the marketplace will expect to be engaged in a completely different way than the field has done it heretofore. Our relevance is at stake. As a sector, we must elevate our standards so that the ability to excel in both heart and head becomes the essential condition of leadership. It is the rare beast who can play well in both KPI (key performance indicator) and EI (emotional intelligence). Speaking for Kendal, that is what we aspire to.

Every element of leadership previously discussed has been tested, exposed, and highlighted by the COVID-19 pandemic, which struck the U.S. in early 2020. We are plainly in unprecedented times. The world is turned upside down and not one person among us has ever dealt with anything like this. Never before has there been so much uncertainty, so much anxiety, so much fear, so much "grey," with little or no black or white.

Yet, despite it all, and despite the prevailing narrative, Kendal and the field by and large have responded to the novel coronavirus in ways that inspire. In these "grey spaces," we are often left to exercise more judgment than ever before and, increasingly, that judgment sits on a foundation of organizational and personal values, values that call us to be transparent, respectful, honest, hopeful, compassionate, wise. There has not been a clear path anywhere in all of this except to recognize the inherent value and strength in every single person among us and at the same time to rely on community and collaboration to yield exceptional actions that have preserved people's comfort, connectedness, and safety in ways one would not have imagined.

The acumen and foresight shown by the leadership across Kendal and in our field has been equaled only by its evident courage and indefatigability. Friends have said to me that this isn't the work they'd "signed-up" for, but the last thing they would do is let down their teams and the residents and members in their care, *no matter what*. For too long, leadership in our sector, as well as our front-line colleagues, have gone unknown or underappreciated. But in the crucible of the pandemic, under extreme scrutiny, they are being seen as the heroes they've always been. We need to carry that message forward to ensure that the world is brought to understand the power in the people that do this incredible work.

From a leadership perspective personally, I have found it ever more important to communicate broadly and intimately. Nothing has been and will be more crucial than frequency and diversity of communications that is reliable, accessible, timely, relevant, soothing, sober, wise, and compassionate.

The subject matter can vary, but a primary goal I have is to sense and coalesce where we need to take positions as a larger organization, to find opportunities and initiatives that we can all get behind. And not just those that identify problems to be solved, but also those that are ends in themselves, which comes from constantly discovering means by which we can feel connected, as one, and a part of this bigger thing that binds us together and gives us strength. We've done that in several areas including testing, work force, discipline around model practices, and standing up for the sector in the face of being cast as a scapegoat.

These things all matter from an internal perspective and in terms of moving the world to a better understanding of what we do and how vital it is. In leadership, it is vital to be able to cast and articulate a vision that the organization can align behind. In times like these, leadership at its best can sense what's most important in the moment and help people be sure that they can accomplish what they need to. Holding people up—that truly is the work, and the joy, in these times.

I have three daughters, two in college, one now moving into middle school. The youngest often reminds me that I always coached her sisters' basketball teams when they were growing up. I didn't coach hers last season, which was the first year that she played travel basketball in our little town, and she reminded me enough to help me realize that I need to be able to make coaching her team a priority in my life.

This is representative. Being present, noticing that something is important to your daughter when they have started to wonder if you see it that way, can change their life, as well as yours. I try to be mindful of applying this thinking to all of my colleagues across our organization and even across our industry. Noticing what is important to them and speaking to it in authentic ways that people can internalize and feel empowered by is so important and so powerful.

This practice connects to so many different things—to care enough about someone or some situation, to notice the specificity of it, to be able to speak to it. If you communicate these moments in ways they can be appreciated, and then connect them to something bigger that others might otherwise not pick up on, then you are on the path to having genuine impact as a leader. Ultimately, leading can come down to being mindful of the incredible people you're lucky enough to be among.

END

Sean Kelly joined Kendal in 2008 and took on his current role as President & CEO in 2016. Prior to 2016, Sean was responsible for fostering a culture of continuous improvement through leading and managing new opportunities for growth and evolution for Kendal. Sean has been a frequent speaker on many topics, including senior housing development, emerging trends, organizational management, leadership, as well as affiliation, merger, and acquisition processes. Sean came to Kendal after 10+ years working with a development, finance, marketing and operations consultant to senior housing and service providers throughout the United States, where he served as Principal. Sean graduated from Ursinus College in Collegeville, PA in 1992 with a bachelor's degree in international relations, and in 2014 he completed an Advanced Management Program at the Wharton School at the University of Pennsylvania in Philadelphia.

GPSR Compliance
The European Union's (EU) General Product Safety Regulation (GPSR) is a set of rules that requires consumer products to be safe and our obligations to ensure this.

If you have any concerns about our products, you can contact us on

ProductSafety@springernature.com

In case Publisher is established outside the EU, the EU authorized representative is:

Springer Nature Customer Service Center GmbH
Europaplatz 3
69115 Heidelberg, Germany

www.ingramcontent.com/pod-product-compliance
Ingram Content Group UK Ltd.
Pitfield, Milton Keynes, MK11 3LW, UK
UKHW021035050925
462611UK00011B/896

* 9 7 8 3 0 3 0 5 3 9 6 8 9 *